Proceedings in Life Sciences

Preceding Danavox Symposia

Published in Scandinavian Audiology

1. Evoked Response Audiometry, 1969, 45 pp.
2. Speech Audiometry, 1970, 266 pp.
3. Electroacoustic Characteristics relevant to Hearing Aids, 1971, 166 pp.
4. Integration of Hearing Handicapped in the Community, 1972, 130 pp.
5. Evaluation of Hearing Handicapped Children, 1973, 269 pp.
6. Visual and Audio-Visual Perception of Speech, 1974, 290 pp.
7. Ear Moulds and Associated Problems, 1975, 307 pp.
8. Sensorineural Hearing Impairment and Hearing Aids, 1977, 480 pp.
9. Auditory Training of Hearing Impaired Pre-School Children, 1978, 112 pp.
10. Binaural Effects in Normal and Impaired Hearing, 1982, 204 pp.

Time Resolution in Auditory Systems

Proceedings of the 11th Danavox Symposium on Hearing
Gamle Avernæs, Denmark, August 28–31, 1984

Edited by
Axel Michelsen

With 112 Figures

Springer-Verlag
Berlin Heidelberg New York Tokyo

Editor:
Professor AXEL MICHELSEN, Institute of Biology, Odense University
5230 Odense M, Denmark

Organizing Committee:
ANNEMARIE SURLYKKE, OLE NÆSBYE LARSEN, LEE A. MILLER, and
AXEL MICHELSEN

ISBN 3-540-15637-2 Springer-Verlag Berlin Heidelberg New York Tokyo
ISBN 0-387-15637-2 Springer-Verlag New York Heidelberg Berlin Tokyo

Library of Congress Cataloging in Publication Data. Danavox Symposium on Hearing (11th: Agernaes,
Denmark). Time resolution in auditory systems. (Proceedings in life sciences). Bibliography: p.
Includes index. 1. Time perception–Congresses. 2. Psychoacoustics–Congresses. 3. Auditory pathways –
Congresses. 4. Temporal integration–Congresses. I. Michelsen, Axel, 1940–   . II. Title. III. Series.
QP445.D36 1984   591.1'825   85-12653.
ISBN 0-387-15637-2 (U.S.)

Offsetprinting and bookbinding: Brühlsche Universitätsdruckerei, Giessen
2131/3130-543210

Preface

Many books from symposia describe the current status
in well established fields of research, where much is
known and where the loose ends are only details in the
picture. The topic dealt with here does not fall into
this pattern. The study of time as a parameter in its
own right is difficult, and the loose ends tend to dominate the present picture. Although the book does
provide the reader with an overview of the field, its
main value is probably to act as a source of "food for
thought" for those interested in the function of sense
organs and nervous systems as substrates for behaviour.

The Introduction is intended to provide the readers of
the book with a short guide to the topics discussed in
the different chapters. The rather detailed Index may
help those looking for information on specific topics.
The Index also explains most of the abbreviations used
in the book.

The basic idea of the Danavox symposia is to invite a
small group of experts to discuss a rather narrow theme
in sound communication. The small number of active participants has the advantage of encouraging intense discussions and of avoiding overloading the program. On
the other hand, selecting the participants is difficult.
The organizers are aware that many other distinguished
colleagues might have been invited. In selecting the
contributors to the 11th Danavox Symposium and thus to
this book we have tried both to obtain a broad and comparative overview of the subject and to provoke in depth
discussions. The discussion during the symposium is not
included as separate sections in the book, but all authors have been asked to modify their contributions to
take into account ideas and arguments brought up during
the symposium. We hope that this procedure has given
the book more continuity.

It is a pleasure for me to thank the Danavox Foundation
for its most generous support. I am very grateful to
my colleagues Annemarie Surlykke, Ole Næsbye Larsen,
and Lee A. Miller for their help and encouragement, to
Hanne Andersen for her careful handling of the manu-
scripts, and to Musse Olesen and Wagn Agergaard for

improving the quality of several of the figures. Finally, I am most grateful to the Springer-Verlag, in particular Dr. Czeschlik and Antonella Cerri for their
understanding and help during the production of the
book.

Odense, 1985   *Axel Michelsen*

Contents

Introduction
A. Michelsen   1

Auditory Processing of Temporal Cues in Insect
Songs: Frequency Domain or Time Domain?
A. Michelsen, O.N. Larsen, and A. Surlykke
(With 15 Figures)   3

Temporal Processing by the Auditory System
of Fishes
R.R. Fay (With 17 Figures)   28

Time Resolution in the Auditory Systems of
Anurans
R.R. Capranica, G.J. Rose, and E.A. Brenowitz
(With 12 Figures)   58

Aspects of the Neural Coding of Time in the
Mammalian Peripheral Auditory System Relevant
to Temporal Resolution
E.F. Evans (With 13 Figures)   74

Theoretical Limits of Time Resolution in Narrow
Band Neurons
D. Menne (With 11 Figures)   96

Time Coding and Periodicity Pitch
G. Langner (With 6 Figures) 108

Temporal Factors in Psychoacoustics
D.M. Green (With 3 Figures) 122

Auditory Time Constants: A Paradox?
E. de Boer (With 2 Figures) 141

Gap Detection in Normal and Impaired Listeners:
The Effect of Level and Frequency
S. Buus and M. Florentine (With 8 Figures) 159

Range Determination by Measuring Time Delays
in Echolocating Bats
H.U. Schnitzler, D. Menne, and H. Hackbarth
(With 9 Figures) 180

Time Constants of Various Parts of the Human
Auditory System and Some of Their Consequences
P.V. Brüel and K. Baden-Kristensen
(With 10 Figures)   205

Temporal Patterning in Speech: The Implications
of Temporal Resolution and Signal-Processing
M. Haggard (With 6 Figures)   215

Subject Index   238

List of Contributors

Baden-Kristensen, K.   205
Brenowitz, E.A.   58
Brüel, P.V.   205
Buus, S.   159
Capranica, R.R.   58
de Boer, E.   141
Evans, E.F.   74
Fay, R.R.   28
Florentine, M.   159
Green, D.M.   122
Hackbarth, H.   180
Haggard, M.   215
Langner, G.   108
Larsen, O.N.   3
Menne, D.   96, 180
Michelsen, A.   1, 3
Rose, G.J.   58
Schnitzler, H.U.   180
Surlykke, A.   3

Introduction

Axel Michelsen

Institute of Biology, Odense University, DK-5230 Odense M, Denmark

The topic of the book is the auditory processing of acoustic signals in the time domain. The scope is restricted to monaural processing, e.i. the binaural processing of minute time cues in directional hearing is excluded. Binaural processing was the subject of the 10th Danavox Symposium. Like most symposium volumes, the book suffers from the weakness that the material on many of the specific subjects is presented and discussed by several contributors. This chapter is intended to provide the readers of the book with a short overview of the problems discussed in the different chapters. The readers are also referred to the Index, both for locating specific subjects and for an explanation of abbreviations.

The problem of defining time resolution and distinguishing it from frequency analysis may be attacked from various angles. One may take advantage of the fact that hearing has evolved independently about 20 times, and that much can be learned about general mechanisms from a <u>comparative approach</u> to the problems. This book contains descriptions of temporal processing in widely different animals: insects (p. 3-27), fishes (p. 28-57), frogs and toads (p. 58-73), and birds (p. 113-119). Most chapters deal with mammals, including humans. Two chapters discuss the remarkable hearing capacity of echo-locating bats (p. 96-107 and p. 180-204). The implications for acoustic instrumentation and for the legislation on noise are discussed on p. 205-214. Finally, temporal processes in the perception of speech are discussed on p. 215-237.

The reception of sound starts with the interaction between a sound wave and the sound-receiving structures (in most cases a tympanum, but see p. 28-30 for a description of the processes in fishes). The <u>mechanical time resolution</u> at this level and in the inner ear reflect the vibrations of membranes and other structures. It has often been assumed that a simple oscillator (a second order system) can be used as a model of the behaviour of these structures. Although this may almost be the case (the moth tympanal membrane, p. 11-12), the use of this simple model for predicting the temporal processing seems to be misleading in most cases (p. 13-15, 74-81, 143-144, 207).

The coding of temporal information in <u>auditory nerves</u> appears to vary. Some animals (e.g. fishes, p. 44-48) seem to be specialized in processing information about frequency in the time domain, whereas in other animals (e.g. insects, p. 15-20, echolocating bats, p. 96-99) the carrier frequency is not reflected in the timing of the nerve impulses. Man and most animals appear to be somewhere between these extremes (p. 48-52, 59-62, 75-81, 111-113).

The processing of temporal information in the <u>central nervous</u>

system has been studied in a wide variety of animals. Specialized neurons tuned to a specific range of amplitude modulation frequency have been found in insects (p. 22-23), fishes (p. 41-43), frogs (p. 65-70), and birds and mammals (p. 115-116).

Not much is known about the cellular mechanisms involved, but evidence for intrinsic oscillations in neurons and coincidence detectors being involved in the temporal processing has been found (p. 108-121).

The thinking about central processing has been much influenced by the concepts of cybernetics. The existence of neuronal auto- and cross-correlating circuits or neurons have often been postulated, and these and other concepts from cybernetics are also used throughout this book (see the Index). The actual use of such processes in neural processing has rarely been the subject of critical scruteny. On p. 187-192 a critical examination of the evidence for such processes in the echolocation of bats is presented.

Much of the knowledge of temporal processing in brains has been obtained from behavioural (psychophysical) experiments. Data from behavioural experiments on animals and humans are discussed by several contributors who have taken advantage of modern methods for creating sound stimuli with well-controlled temporal and spectral properties (p. 48-53, 126-129, 148-154, 163-164). The results of the studies on natural sound signals (insects: p. 5-11, frogs: p. 58-59, human speech: p. 227-234) are often less clear-cut, but the use of computers for modifying such sound signals has now provided some understanding of the amounts of processing in the time and frequency domains during the perception of natural signals. Studies on temporal processing in humans have also provided evidence for pathological cases of temporal processing of (p. 171-177).

The value of an introduction of this kind decreases with its length, and this Introduction therefore is kept very brief. The reader is referred to the Index for further access to specific topics.

Auditory Processing of Temporal Cues in Insect Songs: Frequency Domain or Time Domain?

Axel Michelsen, Ole Næsbye Larsen, and Annemarie Surlykke

Institute of Biology, Odense University, DK-5230 Odense M, Denmark

INTRODUCTION

This contribution is an attempt to answer the questions: 1) Is it possible from the present knowledge of peripheral and central auditory processing to account for the behavioural preferences shown by listening insects? 2) Do we have evidence elucidating whether the processing of rapid changes in the sound signals is done in the time domain, in the frequency domain, or in both domains?

We shall not be able to give clear answers partly because we are not quite sure how to make clear distinctions between time - and frequency domain processing. It might therefore be tempting to remember the saying "Oh Lord, help me to keep my big mouth shut, until I know what I am talking about". On the other hand, by adressing such questions we may encourage ourselves and others to provide some clear thinking and experiments.

Ideally, one should select an experimental animal with clear behavioural preferences in experiments, where minute time cues are being manipulated. One should then study both the peripheral and central processing in this animal in order to account for the behavioural results. At present, we are not able to do this. The best we can do is to pool information from different insects. This approach is not ideal, since a sense of hearing has evolved independently at least a dozen times in different insects.

In the following sections, we consider the evidence for auditory processing in the frequency- and time domains, which has been collected in behavioural, physical, and neurophysiological experiments.

BEHAVIOURAL TIME RESOLUTION

The vast majority of studies of sound communication in insects support the notion that the behavioural information is carried by the gross rhythmicity of the songs. Of course, the carrier frequencies have to be within the range of hearing of the animals, but insects often respond to very crude models of their communication songs. This was first observed around 1910 by the founder of experimental analysis of sound communication in insects, the Slovenian biologist Johann Regen. He succeeded in attracting female crickets to a telephone transmitting the song of a male cricket. Given the quality of transmission of the telephones at that time,

Regen concluded that details in the songs could not be terribly important for the animals (Regen 1913).

Many insects emit sounds composed of long series of short impulses with broad frequency spectra. Shorthorned grasshoppers and bush crickets, for example, may be attracted by white noise models of their songs, provided that the amplitude modulation is about right. It is even possible to persuade a grasshopper to sing in duet with a typewriter! From such observations, it is tempting to conclude that - with a carrier in the audible range of the animals - only gross cues in the time domain are important for the recognition of the songs.

Considering this apparent lack of frequency selectivity, it is puzzling why so many insects have invested in frequency analysers. About twenty years ago it was found that locusts have the ability to perform a frequency analysis of the sound carrier. We now know that most insects using sound for social communication are able to perform a frequency analysis (e.g. crickets, bush crickets, short-horned grasshoppers, mole crickets, various bugs).

One reason for this confusion may be that insects are not very selective in their behavioural responses, unless they are forced to select. A grasshopper which has been socially isolated for some time may respond to many different sounds, but the same animal is probably able to show much more selectivety when surrounded by many different singers in its natural habitat. Recent studies also demonstrate that grasshoppers presented with unnatural imitations of their songs may require a few dB higher sound levels in order to respond (Skovmand and Pedersen 1983, Johansen 1982). It may be, therefore, that some investigators have obtained the false impression of very little discrimination, because the experiments have not been done in the right way. We should like to stress that no really solid evidence is backing this notion.

Gap detection

Psychophysical experiments are not easy to perform in insects, which do not show any interest in the positive or negative rewards normally offered in such experiments. Instead, one may take advantage of their interest in the social signals exchanged during the sexual behaviour. This has been done by Dagmar and Otto von Helversen in a series of studies on shorthorned grasshoppers (1972, 1979, 1983, 1984).

Grasshoppers sing by stridulation, i.e. by hitting a series of cuticular teeth by a scraper. The series of teeth (the file) and the scraper are located on two parts of the body which can be rubbed against each other (in shorthorned grasshoppers the hindlegs and the wings). Each tooth impact gives rise to an impulse vibration (Fig. 1). A series of impulses may be grouped to form a syllable, and two or more syllables may form a chirp, and so on (Fig. 1). The numbers and durations of each of these components differ among different species. To a first approximation these songs look like amplitude-modulated broad-band noise.

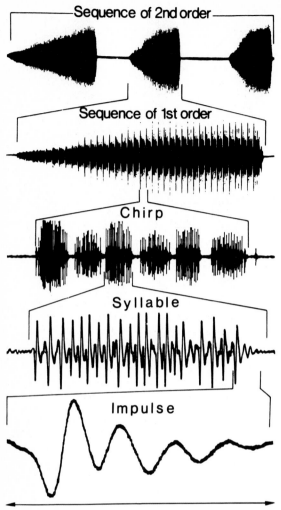

Fig. 1. The structure of the song of a gomphocerine shorthorned grasshopper (Chorthippus biguttulus) and the terminology proposed by Elsner (1974). The time mark (←→): 15 s (sequence of 2nd order); 2.9 s (sequence of 1st order); 72.5 ms (chirp); 9.1 ms (syllable); 0.9 ms (impulse). (From Elsner 1974).

In the grasshopper <u>Chorthippus</u> <u>biguttulus</u> the song consists of syllables of about 10 ms duration (the durations of the components of insect songs depend on temperature). Six syllables form a chirp and about 30-40 chirps form a sequence of 1st order, which may be repeated many times (Fig. 1). Animals in the right mood may respond to "acceptable" artificial songs by singing a response song. Artificial songs are easily made by amplitude modulation of white noise. Both the durations of the components and the pauses between them must be within certain ranges in order for the song to be acceptable.

Of special interest in the present context is that even a very small increase in the pauses between the syllables may cause a dramatic decrease of the acceptability of a song (von Helversen 1972). This is illustrated for a single female in Fig. 2, where the pause between the syllables has been varied at two different syllable durations (3 ms and 10 ms, corresponding to 12-18 and 5-6 syllables per chirp, respectively). For the 3 ms syllables an

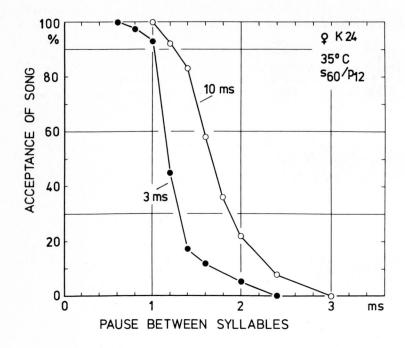

Fig. 2. Response rates of a female shorthorned grasshopper (Chor-thippus biguttulus) to artificial songs with syllable durations of 3 ms and 10 ms, respectively. The pauses between the syllables were varied. The chirp duration and the pause between the chirps were close to the most attractive values: 60 ms and 12 ms, respectively. (From von Helversen 1972).

increase of the pauses between the syllables from 1 ms to 1.2 or 1.4 ms are sufficient for causing a decrease in the response rate to one half and one fifth, respectively. At the more natural syllable duration of 10 ms a similar effect of the pause duration is seen.

Later experiments (von Helversen 1979) have shown that this "gap detection" is a function of the sound level. The gaps between the syllables must be longer at lower sound pressure levels in order to be detected by the listening grasshoppers. At levels about 3 dB, 15 dB, and 27 dB above threshold, the minimum durations of the gaps detected were about 3.6 ms, 1.6 ms, and 1.2 ms, respectively.

These experiments demonstrate a surprising capacity in grasshoppers for detecting minute details in the amplitude modulation of the sound carrier. However, it may be questioned, whether these pause durations should be compared with the gap detection thresholds measured in vertebrates and human beings (Fay 1985; Green 1985; Buus and Florentine 1985). The pauses occur at least 50 times per second, and the listening animals may perhaps take advantage of this redundancy.

Timing of frequency components

In the experiments by the von Helversens the artificial songs were made by amplitude-modulating band-limited white noise. The natural songs have a more complicated fine structure, however, and this fine structure also seems to play a (modest) role in carrying behavioural information.

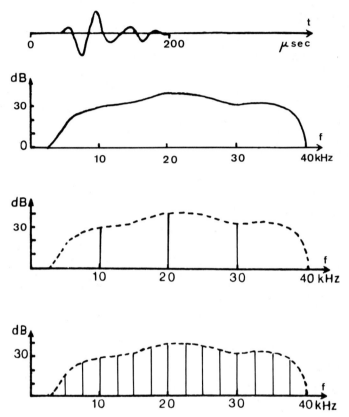

Fig. 3. The spectra of a single impulse (shown in the time domain in A) and of a series of impulses. (B) Single-impulse sounds have their energy rather evenly distributed over a broad frequency range (components above 40 kHz have been removed by filtering). (C), (D) A series of regularly-spaced impulses has the energy concentrated in a few spectral lines. Note the difference between (C) (10.000 impulses s^{-1}) and (D) (2500 impulses s^{-1}). (Redrawn from Skovmand and Pedersen 1978).

The spectrum of the single impulses from each tooth impact in stridulation covers a broad frequency range (Fig. 3B). However, a series of impulses may give rise to a line spectrum if the impulses are regularly spaced. In such a line spectrum the fundamental frequency component is determined by the time interval between the impulses, and the other lines are harmonics of the fundamental one (Fig. 3, C and D). In insect songs, the spacing of the impulses is not so regular that line spectra occur. Instead, the spectrum has a number of maxima corresponding to the fundamental and harmonics of the instantaneous impulse rate.

The spacing of the impulses has been particularly studied in the shorthorned grasshopper <u>Omocestus</u> <u>viridulus</u>, in which each chirp consists of two syllables, one produced by the leg upstroke and one by the downstroke. The impulse rate varies in a systematic way during a chirp from about 250 Hz to about 4 kHz (Fig. 4A). Such

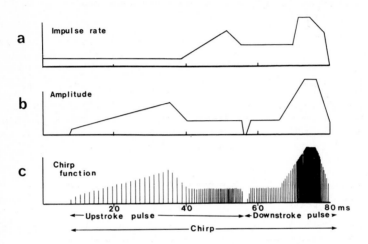

Fig. 4. Both the amplitude of the impulses (corresponding to the force of the individual tooth impacts) and their rate change during the calling song of the shorthorned grasshopper Omocestus viridulus. The figure shows the impulse rate (a) and amplitude function (b) used for creating an artificial song (c) by means of a digital computer. (From Skovmand and Pedersen 1983).

frequency sweeps are present in the songs of several insects. Since the spacing between the impulses is changing with time, the position of the spectral maxima also change during each chirp.

During each chirp, the singing animal changes not only the velocity of its hindlegs (and thus the rate of impacts from the regularly spaced teeth) but also the force between the scraper and the file. The impulses therefore vary both in rate and amplitude during a chirp - and there is also a variation of the emitted sound power. In <u>Omocestus</u> <u>viridulus</u> the emitted sound power reaches a maximum close to the end of each chirp, because both the number of impulses per unit of time and their amplitude are at a maximum (Fig. 4).

In summary, the listening animals are presented not only with a gross amplitude modulation, but also with fast sweeps of both sound power and carrier spectrum. The question now is: do they use this additional information?

This question - whether the listening grasshoppers pay any attention to the fine details in the song - must be answered by means of

behavioural experiments, in which these details are varied.
Ideally, one should perform the analysis in such a way that the
sweeping frequency spectra are varied independently of the sweeping
sound power. Some attempts have been made in recent years to do
this, but an ideal experiment would require a determination of the
time constant for the perception of the sound power. This information is not yet available. Nevertheless, one can learn much by
comparing the "attractiveness" of natural and artificial songs to
listening grasshoppers. The "attractiveness" is assumed to be indicated by the threshold sound level for releasing a response song
from the listening animal.

A large number of interesting artificial songs can be made by means
of digital computers, which also allow for a very detailed signal
analysis. The procedure for creating the artifial songs is to record an impulse sound from a grasshopper song (Fig. 3A), feed it
into the computer, and repeat it with the amplitudes and time
intervals chosen by the programmer. One may also separate different
frequency bands by filtering the natural or artificial songs and
recombine the frequency components with slightly changed time relations between the components (Skovmand and Pedersen 1978 and 1983,
Johansen 1982).

Without going into the technical details we may note that even very
slight modifications of a natural song may cause a decrease in
"attractiveness" (some dB increase in the threshold for releasing a
behavioural response). Most artificial songs are less attractive
than the natural ones, but some are just as attractive ("super-
optimal" artificial songs have not been found). At present, it is
not quite obvious how these results should be interpreted, but it
is noteworthy that both attractive and less attractive songs have
the same gross amplitude modulation. The frequency spectrum and
sound power is also the same for all songs when measured with time
constants corresponding to the duration of syllables or chirps. The
songs differ, however, when these parameters are measured with
shorter time constants.

In order to interpret the behavioural results, one must know the
exact position of the frequency analysing "filters" in the peripheral and central auditory pathways of the species studied, and
one also has to know the "time constants" for the response of these
neurons to sudden changes in the sound power within their frequency
ranges. Although these properties are not known, the present knowledge about the physiology can be used for formulating a working
hypothesis for the analysis of the songs in Omocestus viridulus.

The ears of shorthorned grasshoppers contain four groups of receptor cells with different frequency sensitivities (Michelsen 1971,
Römer 1976). In the locust, three of the cell groups (named a-c)
have their best frequencies between two and eight kHz, whereas the
fourth group (named d) is mainly sensitive to sounds above 10 kHz
(Fig. 5). The exact position of the threshold curves is not yet
known in Omocestus viridulus, but whole-nerve recordings suggest
that the a-c cells may have best frequencies slightly higher than
in the locust. With the computer, one can create digital frequency
filters simulating the receptor cells. These filters can then be
used for studying the timing of the energy flow to each group of

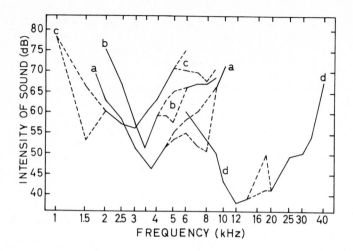

Fig. 5. The ear of shorthorned grasshoppers and locusts contains four groups of receptor cells (called a, b, c, and d) with different frequency sensitivities. The figure shows the threshold curves of receptor cells in an isolated ear preparation (Michelsen 1971). According to Römer (1976) the intact ear has rather similar frequency-analysing properties.

receptor cells.

The frequency analysis in the ears of locusts and other shorthorned grasshoppers is caused by the tympanum and the cuticular attachment bodies of the receptor cells acting as a mechanical frequency analyser, the detailed function of which is still debated. As already mentioned, single tooth-impact impulses cover a broad frequency range (Fig. 3B), and they therefore activate all four groups of receptor cells in the ear. The same is true for series of impulses in the part of the chirps, where the impulses are widely spaced. During the fast leg movements, however, the position of the maxima in the frequency spectrum may shift rapidly. The high-frequency d-cells have so broad threshold curves that they will always be activated, but the three groups of low-frequency cells are more narrowly tuned. Let us consider a group of cells tuned to 4 kHz. These cells will be maximally stimulated when the tooth impact frequency is around 2 kHz (by the first harmonic component of the spectrum) and again around 4 kHz, but they will receive less energy at other tooth impact frequencies. The three groups of receptor cells sensitive to sounds below 10 kHz will all be activated during the most intense part of each chirp, where the impulse rate increases to about 4 kHz, but not always simultaneously.

The preliminary analysis of the natural and artificial songs used in the behavioural experiments suggests that the animals prefer songs, which cause an almost simultaneous activation (within a time interval of about one millisecond) of the different groups of receptor cells during the maximum close to the end of each chirp (Fig. 4). Less attractive songs tend to cause a larger scatter in times for maximum activation of the groups. It should be stressed, however, that the timing of activation of the receptor cells was inferred from a model: the computer with its digital filters (Johansen 1982). Recordings from single receptor cells in the four groups would be much more convincing.

Conclusion and discussion

Although there are several loose ends in this story, the experimental results suggest that some insects may be sensitive to time cues of the order of a few milliseconds. The fact that white noise was used for the song models tested in the "gap detection" experiments may suggest that the neural processing was done exclusively in the time domain. In the experiments with <u>Omocestus</u>, however, it is more likely that the song recognition is based partly on the "correct" timing of the peak activities of different groups of receptor cells responding to different frequency bands, i.e. that the neural processing is carried out both in the frequency- and time domains.

In this context, it is interesting that singing grasshoppers do not make the amplitude pattern as clear as possible. In contrast, the fine patterns made by each of the animal's two "intruments" (the two pairs of hindlegs and wings) are produced somewhat out of phase, so the pauses between syllables are camouflaged. The resulting "blurred" output is not caused by sloppy coordination of the legs, since the phase shift between the legs is timed with a precision of a millisecond (Elsner 1974). Really nice songs with a clear amplitude pattern are produced only by invalid grasshoppers lacking one of their hindlegs, but there is no evidence that the songs of these animals are more attractive than normal songs. That "blurred" songs are attractive is understandable, if the fine structure is not perceived directly in the time domain, but only after a preliminary analysis in the frequency domain.

MECHANICAL TIME RESOLUTION

A convenient way of investigating the reactions of a system to fast changes in its input is to evoke its impulse response. This can be done by activating the system with a short, intense force (known as Dirac's Delta function). In simple mechanical systems (e.g. a mass connected to a lightly damped spring) the impulse response may have the form of a regularly decaying oscillation. The time constant of the decay can be regarded as a measure of the "memory" of the system. If the system is linear, the impulse response can be used for predicting the reactions to all other inputs.

The impulse response of a tympanal membrane vibrating in its basic mode may be rather close to this simple case. The time resolution of the tympanum can be illustrated by activating the ear with a pair of short impulse sounds. This is illustrated in Fig. 6 for the tympanal membrane of a noctuid moth, in which the vibrations were measured with laser vibrometry (Schiolten et al 1981). The best frequency of the tympanum was about 25 kHz, and each of the two sound impulses had a duration of 16 µs and a uniform amplitude

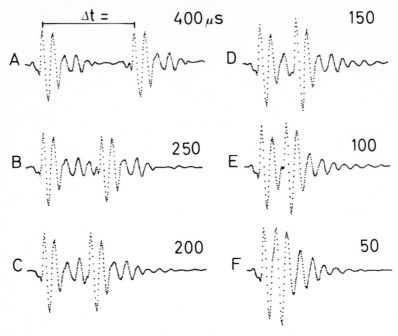

Fig. 6. *The time resolution in the tympanum of noctuid moths is investigated by observing the vibrational response (measured with laser vibrometry) to a pair of 16-μs acoustic impulses with impulse intervals (Δt) varying from 400 μs (in A) to 50 μs (in F). (From Schiølten et al. 1981).*

spectrum between 18 and 38 kHz. It is seen that the impulse responses of the tympanum are separated when the time interval between the impulse sounds is larger than 300-400 μs; that visually separate impulse responses are obtained at intervals down to 150-200 μs; and that the impulse responses superimpose to form one damped vibration when the time interval is smaller than 100 μs.

Such direct observations of a system's behaviour in the time domain are not easy to perform. Ideally, the duration of the Delta-function should be less than 10% of the time constant of the system under study (Varjú 1977). The time constant of the moth tympanum is about 60 μs, so ideally the impulse sound should have a duration of less than 6 μs. Since this is not the case, the impulse responses are somewhat distorted. In practice, it has proven extremely difficult to produce intense impulse sounds with ideal properties: short duration, reproducible and variable amplitude, and a flat amplitude spectrum within the audible range. For example, some investigators have used electric sparks. Although single sparks may have properties close to the ideal ones, they are seldomly reproducible. It is also difficult to control the timing of spark generation. The only obvious advantage of sparks seems to be that no triggering wires are required: all computers and other instruments in the neighbourhood trigger automatically! Our present favourite method of creating acoustic Delta-functions is to let a very thin plastic film jump from one position to another. The 15 μs impulses thus produced have flat amplitude and phase spectra from about 800 Hz to about 80 kHz. Although these impulses are well suited for activating ears within the frequency range mentioned, they are not suited for studies below 500 Hz (the energy is too small at low frequencies). A method for producing acoustic Delta-functions with sufficient energy also at low frequencies would be most welcome.

The practical problems involved with the direct observations of the impulse responses may be circumvented. One may either determine the transfer function of the system (e.g. by using pure tones or noise as stimulus) and then calculate the impulse response by applying the inverse Fourier transform. Alternatively, one may perform a cross-correlation between the input (white noise sounds) and the output (e.g. the vibration of the tympanum) and thus obtain the impulse response. In theory and for linear systems these kinds of approach should be quite as reliable as the direct observation. In practice, however, it is often difficult to be sure that the impulse responses thus obtained are fair representations of the real impulse response (Schiolten et al. 1981). This seems especially to be the case for complicated systems like transmission lines (e.g. the cochlea).

Although the impulse response of a simple tympanum like that of the noctuid moth looks much like the impulse response from a second order system (e.g. a mass connected to a spring), the oscillatory decay is more complicated in the tympanum. The use of a time constant for describing the decay is therefore only an approximation. Much more complicated impulse responses are found in the tympanal membranes of most animals and man (Fig. 7). Some of these impulse responses are so complicated that the application of a single time constant for characterising the decay is not appropriate.

The reasons for the different shapes of the impulse response are diverse. In many animals (e.g. most insects, frogs, lizards, and even some birds and mammals) the tympanal membrane receives sound not only at its outer surface, but also at the inner surface, because they need their ears to work as pressure difference receivers in order to perceive the direction of the sound waves (see e.g. Michelsen 1983). Sound is therefore allowed to reach the two surfaces of the tympanum through different paths. A short sound impulse may therefore hit first one side and then the other side of the tympanum, thus creating an impulse response of complicated shape (Larsen 1981). This type of complexity is not reflecting the mechanics of the tympanum, but the transfer function of the acoustical parts of the hearing organ. An example of such complicated acoustics is found in crickets, where each ear has no less than four acoustic inputs (Larsen and Michelsen 1978). It should be noted that the time of arrival of the different sound waves reaching the tympanum depends on the direction to the sound source. The shape of the impulse response is therefore a function of the direction of sound (Fig. 7A and B).

The complicated shape of the impulse response may thus reflect the complicated mechanics of the tympanum and/or other components of the ear. The impulse response of human tympana consists of at least two phases, a fast response and a slow decay (Fig. 7C). Some tympanal membranes like those of shorthorned grasshoppers and

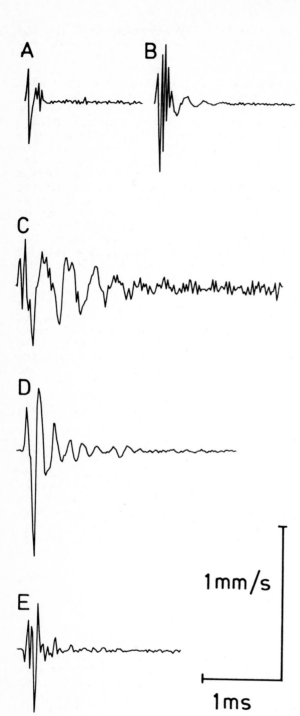

Fig. 7. Impulse responses (i.r.) evoked by short (about 16 μs) sound impulses of moderate intensity (spectral power about 70 dB SPL). A and B: i.r. of a cricket tympanum evoked by a frontal (A) and an ipsilateral (B) sound source, respectively. C: the i.r. of a human tympanum consists of at least two components. D and E: i.r. from a locust tympanum (attachment area of the b-receptor cells) evoked by sound impulses with different power spectra (main energy around 4 kHz (D) and 17 kHz (E), respectively). In all cases, the sound impulses contained only little energy below 800 Hz, and the i.r.s were low-pass filtered at 25 kHz.

1mm/s

1ms

locusts perform vibrations not only in the basic mode, but also at higher modes (Michelsen 1971), and the mode of decay reflects the complexity of the mechanics.

In such complicated systems, the impulse response may depend rather much on the power spectrum of the acoustic impulse evoking them. The term impulse response should therefore be used only for the response to a very short stimulus with flat amplitude- and phase-spectra within the frequency range considered. This is illustrated for the locust ear in Fig. 7D and E. Obviously, in this case the time constants are considerably shorter for the higher (high frequency) modes of vibration than for the lower ones. The reason probably is that in locust ears the main source of resistive damping is in the tympanal membrane and the receptor organ (Michelsen 1971). A certain amount of damping will cause the oscillation to decay over approximately the same number of cycles at both basic and higher modes of vibration. In other words: if it takes, for example, four cycles to reach 20% of the original amplitude both at 4 kHz and at 20 kHz, then the time constant is a factor of 5 smaller at 20 kHz than at 4 kHz. It is important to realize, however, that one may find just the opposite frequency dependence in other ears if the main resistive damping is caused by the interaction with the medium (because the radiation of sound - and thus loss of energy - is less efficient when the tympanum vibrates in a higher mode). Alternatively, the frictional losses (per cycle of vibration) may vary with frequency. At present, such thoughts are only speculations, and this problem should be studied further in various ears.

Conclusion

Let us now return to the problems outlined in the previous section and compare the mechanical and behavioural time resolution in short-horned grasshoppers. Assuming a time constant for the behaviour of 1 ms, the mechanics of the ears is not likely to limit the total time resolution at high frequencies (Fig. 7E, see also Fig. 5 in Schiolten et al. 1981). At low frequencies, however, it is not certain that the mechanics can be neglected (Fig. 7D).

NEURAL TIME RESOLUTION

Receptor responses

Information about the ability of insect auditory receptor cells and auditory neurons to respond to rapid changes of sound power may be found scattered in the literature, but no real quantitative analysis exists for any insect ear. Several authors give some indication of the ability of the cells to synchronize their responses to the clicks in a series of clicks, but the spectral and temporal properties of the clicks are seldomly described. These studies are therefore less useful.

Most hearing insects possess 1-80 auditory receptor cells. The cicadas, however, boast a total of up to 2200 primary fibres in each ear (Doolan and Young 1981). This large number of receptor cells might conceivably be an adaptation to high temporal resolution. Assuming the properties of a perfect parallel receiver the coding precision of time parameters would be increased by a factor of 47 (= $2200^{1/2}$) over that of a single neuron. Unfortunately, studies on this possibility have not been done. Neither is it known whether precise information on time parameters has any behavioural significance in cicadas.

In the migratory locust (a shorthorned grasshopper), the temporal resolution of the four groups of receptor cells has recently been studied by a system-theory approach (Sippel and Breckow 1983). White noise and sinusoids were used for amplitude modulation of the carrier frequency which was chosen to be the best frequency of the group under study. The results were used for calculating the linear and one non-linear component of the transfer function. Apparently, the non-linear component is not very important under the conditions studied (the experimental data fit a linear model almost as well as a combined linear and non-linear model). Unfortunately, this investigation did not contribute much to solve the problem of the limits for time resolution, because the use of a Hann-window with a frequency limit of 100 Hz for filtering the neural response histograms caused the power spectra of the transfer functions to decrease strongly above 100 Hz. It would be interesting to have these experiments repeated without this restriction.

In our laboratory, we have recently collected some data on the ability of the receptor cells in moth ears to synchronize their responses with amplitude modulations of stimulus sounds (Surlykke et al. in prep.).

Moths are especially well suited for studies of auditory transduction processes, since the ears are very simple: one or two receptor cells attach to the centre of a fairly homogeneous tympanum, the mechanical behaviour of which has been described above.

In the experiments 40 kHz pure tones and band limited noise sinusoidally modulated between 10 and 400 Hz served as stimuli. The best frequencies of the auditory receptor cells vary between 20 and 50 kHz in different species of moths. The ears are rather broadly tuned (Q_{3dB} about 3), and the receptor cells are therefore unlikely to detect the minute changes in the frequency spectrum of the stimulus sounds caused by the 10–400 Hz sinusoidal amplitude modulations. The synchronization observed between the low-frequency sinusoid and the occurrence of the spikes therefore reflects processes in the time domain (Fig. 8).

The sound level was varied up to 20 dB above threshold and the modulation depth was varied between 0% (no amplitude modulation) and 100% (very brief silent intervals between each carrier pulse).

Various methods may be used for calculating the degree of synchronization of the spikes to the modulating sinusoid. We recorded the spike responses on a signal averager as PST histograms. The signal averager performed an integration of the histograms, i.e. displayed

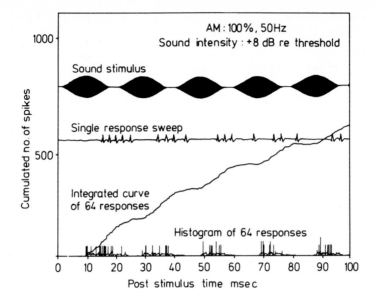

Fig. 8. The time resolution of the receptor cells in a moth ear analysed by estimating the degree of synchronization of spike responses to the amplitude modulation of the sound stimulus. The single response sweep shows that the spikes are phase locked to the maximum amplitude of the stimulus with about 5 ms latency. The PST histogram of 64 consecutive stimulations 8 dB above threshold (lower curve) and the cumulative curve resulting from integrating the histogram both show a clear synchronization to the stimulus.

a cumulative response curve. The time window of the signal averager was chosen to include 3-5 cycles of the sinusoid (3-5 peaks and valleys in the histogram - Fig. 8), and the integration of the synchronized response produced a cumulative curve with 3-5 steps (Fig. 8). Each peak in the original histogram corresponds to a maximum slope in the cumulative curve (the response-slope) and each valley to a minimum slope (the pause-slope). The ratio between tangent to the response-slope and tan (pause-slope) may then be used as a measure of the degree of synchronization. In theory, the ratio may vary between 1 (no synchronization) and ∞ (perfect synchronization). In the experiments we obtained values between 1 and about 100.

This simple method allows the degree of synchronization to be measured and compared at different modulation depths, modulation frequencies, and sound levels above threshold. It is probably not a very sensitive indicator of the threshold for the detection of amplitude modulation (leaving it to the imagination of the investigator, whether an almost straight cumulative curve really is straight). One may argue, however, that thresholds are arbitrary, since the threshold of the central nervous system is unlikely to correspond to that selected by the investigator.

The response synchronization was proportional to modulation depth at all modulation frequencies, and at all stimulus intensities.

At a given modulation depth, the response depends on the modulation frequency and intensity of the stimulus sound. At high intensities the spikes are so regularly spaced immediately after stimulus onset, that this may give a false impression of synchronization: the cumulative curve is staircase-like (Fig. 9). At low inten-

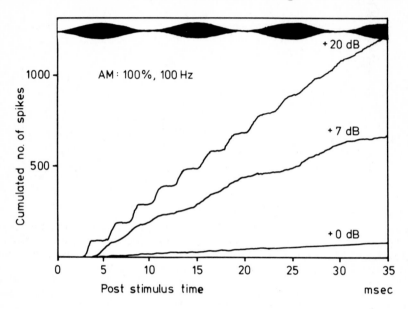

Fig. 9. The influence of sound level on synchronization immediately after stimulus onset. The cumulative curves (cf. Fig. 8) to a 100 Hz, 100% modulated 40 kHz pure tone (upper curve) are shown at three sound levels (0 dB corresponds to the threshold of the receptor cell).

sities, just above threshold, the cumulative curve is a straight line (no synchronization). But at sound levels of 5 to 10 dB above threshold, the time course of excitation and adaptation allow the receptor cells to respond with maximum phase locking to the sinusoidal amplitude modulation of the sound (Fig. 9). Immediately after stimulus onset the synchronization is rather low, because the spike rate is high, but after 30-100 ms the synchronization is a maximum, then dropping progressively with stimulus length (Fig. 10). Hence, at ideal conditions (i.e. 100% modulation depth, sound level 5-10 dB above threshold, and 30-100 ms after the onset of sound) the receptor cells are able to follow amplitude modulations of about 300 Hz (Fig. 11).

In the steady-state, i.e., when using continuous stimulation, the cells can only follow modulation frequencies up to about 150 Hz. At frequencies below 150 Hz, the degree of synchronization is a function of sound intensity - at moderate sound levels. Above 150 Hz, the results are similar to those described above: increasing sound intensity above a critical range does not increase synchronization. These results seem to be rather similar to those

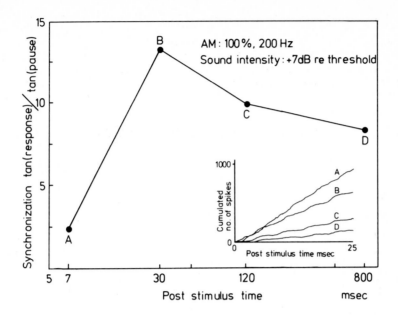

Fig. 10. The degree of synchronization in a moth ear as a function of the time elapsed from the onset of the 1000 ms stimulus. Note the logarithmic scale of the abcissa. Inset: the actual cumulative curves from which the synchronization values were calculated.

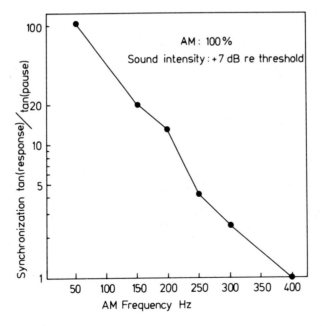

Fig. 11. The degree of synchronization in a moth ear as a function of modulation frequency.

obtained by others on vertebrate preparations (Capranica 1985, Fay 1985).

In contrast to the situation in some vertebrate studies (e.g. Fay 1985, Fig. 10), we do not find any difference between the responses to pure tone and to noise carriers. This is not surprising, since the carriers and modulation frequencies are far apart, and because the change in the frequency spectrum caused by the modulation is unlikely to be detected by the receptor cells (see above). It would be interesting to persue this in different animals in order to elucidate the possible contribution of frequency analysis to the different responses to pure tone and noise carriers observed in some animals.

Central coding of slow amplitude modulations

At the level of the central nervous system of insects only few studies have been undertaken to investigate the abilities of neurons for coding temporal patterns, and all studies concern the coding at amplitude modulation frequencies below about 100 Hz. Such slow amplitude modulations occur in the songs of crickets. The calling songs in Figure 12 are from a cricket habitat in Azerbeidjan in USSR. Six sympatric cricket species are calling together. Each song has an almost pure carrier close to 5 kHz (the higher harmonics are 20-40 dB below the level of the fundamental). The sounds are emitted as a train of syllables forming a chirp (cf. Fig. 1). All amplitude modulations are slow - in the range of 30-100 Hz. The calling songs of the six species differ little in their frequency spectra, but somewhat more in their slow temporal patterns. Since the crickets are obviously adapted to process such simple sounds, it should be relatively easy to investigate and account for temporal resolution in these animals.

The relative importance of the carrier frequency and the temporal pattern can be determined in behavioural tests, where female crickets may express their preferences by running or not running towards the sound source. Such behavioural tests do not reveal any high degree of specificity as regards carrier frequency. Carriers in the range 4-5 kHz are most attractive, but carriers in the range from 2.5 kHz to 12 kHz may elicit tracking of the sound source (Hill 1974, Oldfield 1980). The syllable repetition frequency is very important, however. In the experiments illustrated by the stimulus paradigms and phonotactic response (hatched area) in Fig. 14, females of Gryllus campestris were presented with artificial songs with a carrier of 5 kHz and a duty cycle of 50%. These songs differ both in syllable repetition frequency and in the number of syllables per chirp, but previous experiments have shown that the latter parameter is of little importance in this species (as long as 3 or more syllables are present in each chirp). The behavioural preference of the crickets (expressed as seconds of tracking per minute of trial) shows bandpass properties with maximum tracking between 20 and 40 Hz, and no tracking at all below 10 Hz or above about 60 Hz. The range of maximally attractive pulse repetition rates correlates well with the natural range of the songs for the cricket species investigated, which at $12^{o}C$ is 20 Hz and at $31^{o}C$ is 45 Hz (Thorson et al. 1982).

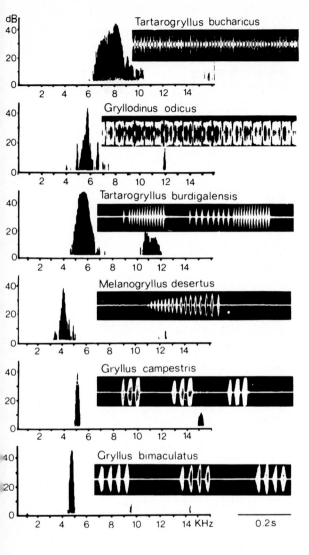

Fig. 12. Frequency spectra and oscillograms of the calling songs of 6 sympatric cricket species from the southern regions of Azerbeidjan, USSR (From Elsner and Popov 1978).

Which mechanisms are responsible for this behavioural preference? Frequency cues are unlikely, since the slight spectral differences between the songs are much too small and within the filter bandwidth (critical band) determined in behavioural experiments on cricket frequency resolution (Ehret et al. 1982). The mechanisms responsible for the temporal tuning are not situated in the ears (which in crickets are in the fore legs). The tympanal membranes react very fast (Fig. 7, Larsen 1981), and the primary receptor cells faithfully copy the pulse repetition at these low rates. The first hint of temporal processing is found in auditory interneurons within or ascending from the prothoracic ganglion. The final temporal tuning is in neurons of the brain.

Wohlers and Huber (1982) noted that the first order auditory interneurons may be separated in two classes according to their time resolution capabilities when tested with the natural song: those that code each syllable with a distinct spike burst (AN1 in Fig. 13), and those that code the duration of the chirp, but not

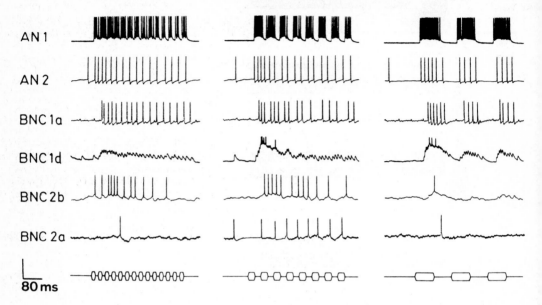

AN 1

AN 2

BNC 1a

BNC 1d

BNC 2b

BNC 2a

80 ms

Fig. 13. Coding properties of central auditory interneurons of the cricket Gryllus bimaculatus in response to simulated calling songs with syllable repetition intervals of 18 ms (lower trace left), 34 ms (middle), and 98 ms (right). AN 1 and 2: ascending neurons, BNC a and d: brain neurons class 1, BNC2 a and b: brain neurons class 2. For further explanation see text (Redrawn from Schildberger 1984b).

consistently the single syllables (AN2 in Fig. 13). A closer analysis showed that in the non-coding neurons the decay-time of the membrane potential was considerably longer than in the coding neurons. If the syllable repetition rate was lowered, even the "non-coding" AN2 neuron would now code for the temporal pattern (Fig. 13, right). One of the two ascending neurons (AN1) will thus code for a wide range of pulse repetition frequencies, while the other neuron (AN2) and the brain neurons BNC1 a and d (Fig. 13) will code only below a certain pulse repetition frequency, i.e. these neurons excibit low-pass properties. Other brain neurons (e.g. BNC2b in Fig. 13) show various degrees of synchronization at high pulse repetition rates but not at low ones, i.e. they have high-pass properties. A third class of auditory brain neurons (e.g. BNC2a in Fig. 13) show band-pass properties, only coding for syllables with repetition rates in a particular range (Schildberger 1984a).

The mechanism producing the high-pass characteristic is not known, but it may be a long lasting neural inhibition that builds up during a long syllable and persists over the next syllable period, while at shorter syllable lengths and repetition rates the excitation dominates the inhibition. The low-pass properties may result from the temporal summation of graded potentials (see BNC 1d in Fig. 13), which at long syllables have time enough to become so large as to generate action potentials, while remaining below threshold level at shorter syllable lengths. The band-pass neurons could obtain their filter characteristics by an ANDing of high- and low-pass filter neurons. Latency measurements and anatomy suggest that this is indeed the case for most of the band-pass neurons. The filter characteristics are summarized in Figure 14, which shows

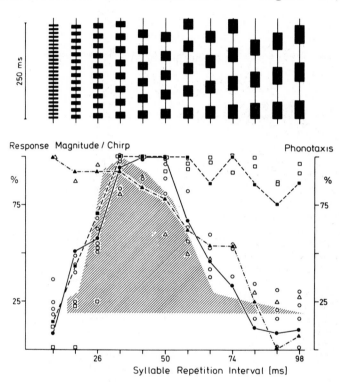

Fig. 14. Responses of auditory brain neurons of the cricket to 250 ms chirps with different syllable repetition intervals, but constant duty cycle (upper part of the figure). In the graph (lower part of the figure) three types of filter characteristics can be distinguished: low-pass filter properties (indicated by triangles, BNC2b), high-pass filter properties (indicated by squares, BNC1d), and band-pass filter properties (circles, BNC2a).
The hatched area shows the relative effectiveness of identical stimuli for eliciting phonotactic tracking (right ordinate, replotted from Thorson et al. 1982). (From Schildberger 1984a).

that not only do low- and high-pass neurons probably add to produce the response of the band-pass neurons, but their band-pass characteristics is closely correlated with that of the animals, when the phonotactic behaviour is measured with identical stimulus paradigms (the hatched area).

Though an auditory interneuron may code for the conspecific amplitude modulations at low sound levels, there is a potential danger of response saturation and hence a cessation of temporal coding at high sound levels. How is the temporal coding of slow amplitude modulations preserved at high sound levels?

It was mentioned that the calling songs emitted by some crickets are carried by rather pure tones. In other species, however, the spectra show strong harmonic components, and the first and second harmonic may be only a few dB below the level of the fundamental which is at 4-5 kHz. Figure 15 illustrates this for three different

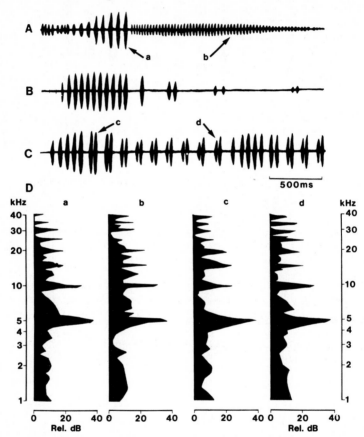

Fig. 15. Oscillograms and frequency spectra of natural songs in the cricket Teleogryllus oceanicus. A. The courtship song, B. The aggression song, C. The calling song, D. Frequency spectra of the syllables marked a-d. (Redrawn from Hutchings and Lewis 1984).

songs in Teleogryllus oceanicus. At low sound levels the fundamental will dominate and interneurons sensitive to this frequency will code the song syllables. At high sound levels the higher harmonics add to the overall sound level stimulating the interneurons, and a blurring of the temporal coding would be expected. However, some interneurons tuned to frequencies in the range of 10 to 30 kHz faithfully code the syllables of the natural songs over a wide intensity range. The two-tone suppression responsible for this ability was elucidated by Hutchings and Lewis (1984) who stimulated such neurons with an 18 kHz carrier modulated in the pattern of the natural songs and found that the discharge of the interneuron was so heavy that syllable intervals were obscured. The addition of a 4.5 kHz carrier re-established the temporal coding abilities - even at the fastest occurring (40-45 Hz) sound pulses. By this two-tone stimulation, temporal coding could be achieved even at sound levels of 98 dB SPL. In classical two-tone suppression experiments, it could be shown that by neural inhibition the frequency components at 4-5 kHz diminish the response to 25 kHz, even when the 4-5 kHz components are at a 20 dB lower sound level. Hutchings and Lewis (1984) concluded that the strong inhibitory side band serves to keep neural firing in the middle of the neuron's dynamic range, ir- respective of the overall stimulus intensity, thus greatly ex- tending the range of coding for details of the temporal pattern.

Two-tone suppression effects are not restricted to single, spe- cialized neurons, but according to Boyd et al. (1984) two-tone suppression by neural inhibition is found in all auditory inter- neurons of crickets - although with different susceptibility to this effect.

More generally, a mechanism of two different frequency regimes interacting to enhance temporal coding seems to be a quite common phenomenon in orthopteran insects. In bushcrickets for instance, the song typically consists of very short impulses forming 15-20 ms syllables, which are repeated at a rate of about 30 Hz for tens of seconds. The auditory interneurons code for the single syllables, while the fine structure of the syllable doesn't seem to convey information. So, again we are talking about rather slow phenomena. All the interneurons show various degrees of habituation, the most strongly habituating ones responding with a single spike only at the beginning of each chirp. It is interesting that some neurons habituating to sound stimuli, dishabituate when the animal ad- ditionally receives input to its low-frequency sensitive vibration receptors (Kalmring et al. 1983).

CONCLUSION

The time resolution demonstrated during the exchange of social signals varies from a behavioural preference for amplitude modula- tions around 30 Hz in crickets and bush crickets to preferences for time cues of the order of one to a few ms in grasshoppers. The former behavioural preference is probably due to auditory proces- sing entirely in the time domain, whereas frequency cues may be involved in the latter case. The mechanical time resolution in the ears does not play any role for the temporal processing in crickets. In grasshoppers, the mechanics is unlikely to limit the time resolution at high carrier frequencies, but it may perhaps be important at low carrier frequencies. The time resolution in peri- pheral and central neurons is sufficient to account for the be- haviour of crickets, but the behavioural preferences for time cues in the ms range in grasshoppers cannot be understood from our present knowledge of the processing in the auditory pathways in insects.

REFERENCES

Boyd P, Kühne R, Silver S, Lewis B (1984) Two-tone suppression and song coding by ascending neurones in the cricket Gryllus campestris L. J Comp Physiol A 154:423-430

Buus S, Florentine M (1985) Gap detection in normal and impaired listeners: The effect of level and frequency. This volume

Doolan JM, Young D (1981) The organization of the auditory organ of the bladder cicada, Cystosoma saundersii. Phil Trans Roy Soc Lond. Ser B 291:525-540

Ehret G, Moffat AJM, Tautz J (1982) Behavioral determination of frequency resolution in the ear of the cricket, Teleogryllus oceanicus. J Comp Physiol 148:237-244

Elsner N (1974) Neuroethology of sound production in gomphocerine grasshoppers. I. Song patterns and stridulatory movements. J Comp Physiol 88:417-428

Elsner N, Popov AV (1978) Neuroethology of acoustic communication. Adv Insect Physiol 13:229-355

Fay RR (1985) Temporal processing by the auditory system of fishes. This volume

Green DM (1985) Temporal factors in psychoacoustics. This volume

Helversen D von (1972) Gesang des Männchens und Lautschema des Weibschens bei der Feldheuschrecke Chorthippus biguttulus (Orthoptera, Acrididae). J Comp Physiol 81:381-422

Helversen D von (1984) Parallel processing in auditory pattern recognition and directional analysis by the grasshopper Chorthippus biguttulus L. (Acrididae). J Comp Physiol A 154:837-846

Helversen D von, Helversen O von (1983) Species recognition and acoustic localization in acrid grasshoppers: A behavioral approach. In: Neuroethology and Behavioral Physiology (Huber F, Markl H, eds.). Springer-Verlag, Berlin, Heidelberg, N.Y. pp 95-107

Helversen O von (1979) Angeborenes Erkennen akustischer Schlüsselreize. Verh Dtsch Zool Ges 1979:42-59

Hill KG (1974) Carrier frequency as a factor in phonotactic behaviour of female crickets (Teleogryllus commodus). J Comp Physiol 93:7-18

Hutchings M, Lewis B (1984) The role of two-tone suppression in song coding by ventral cord neurones in the cricket Teleogryllus oceanicus (Le Guillou). J Comp Physiol A 154:103-112

Johansen MH (1982) Undersøgelse af signalbærende parametre ved den akustiske kommunikation hos markgræshopper. Masters thesis. Odense University

Kalmring K, Kühne R, Lewis B (1983) The acoustic behaviour of the bushcricket Tettigonia cantans. III. Co-processing of auditory and vibratory information in the central nervous system. Behav Process 8:213-228

Larsen ON (1981) Mechanical time resolution in some insect ears. II. Impulse sound transmission in acoustic tracheal tubes. J Comp Physiol 143:297-304

Larsen ON, Michelsen A (1978) Biophysics of the ensiferan ear. III. The cricket ear as a four-input system. J Comp Physiol 123:217-227

Michelsen A (1971) The physiology of the locust ear. Z vergl Physiol 71:49-128

Michelsen A (1983) Biophysical basis of sound communication. In: Bioacoustics. A comparative Approach. (Lewis B, ed.). Academic Press, London, pp 3-38

Oldfield BP (1980) Accuracy of orientation in female crickets, Teleogryllus oceanicus (Gryllidae): Dependence on song spectrum. J Comp Physiol 141:93-99

Regen J (1913) Über die Anlockung des Weibchens von Gryllus campestris L. durch telephonisch übertragene Stridulationslaute des Männchens. Pflügers Arch 155:193-200

Römer H (1976) Die Informationsverarbeitung tympanaler Rezeptorelemente von Locusta migratoria (Arcrididae, Orthoptera). J Comp Physiol 109:101-122

Schildberger K (1984a) Temporal selectivity of identified auditory neurons in the cricket brain. J Comp Physiol A 155:171-185

Schildberger K (1984b) Recognition of temporal patterns by identified auditory neurons in the cricket brain. In: Acoustic and vibrational communication in insects (Kalmring K and Elsner N. eds.) Paul Parey Verlag

Schiolten P, Larsen ON, Michelsen A (1981) Mechanical time resolution in some insect ears. I. Impulse responses and time constants. J Comp Physiol 143:289-295

Sippel M, Breckow J (1983) Non-linear analysis of the transmission of signals in the auditory system of the migratory locust Locusta migratoria. Biol Cybern 46:197-205

Skovmand O, Pedersen SB (1978) Tooth impact rate in the song of a shorthorned grasshopper: A parameter carrying specific behavioral information. J Comp Physiol 124:27-36

Skovmand O, Pedersen SB (1983) Song recognition and song pattern in a shorthorned grasshopper. J Comp Physiol 153:393-401

Thorson J, Weber T, Huber F (1982) Auditory behaviour of the cricket. III Simplicity of calling-song recognition in Gryllus, and anomalous phonotaxis at abnormal carrier frequencies. J Comp Physiol 146:361-378

Varjú D (1977) Systemtheorie für Biologen und Mediziner. Springer-Verlag, Berlin, Heidelberg, New York

Wohlers DW, Huber F (1982) Processing of sound signals by six types of neurons in the prothoracic ganglion of the cricket, Gryllus campestris L. J Comp Physiol 146:161-173

Temporal Processing by the Auditory System of Fishes

Richard R. Fay

Department of Psychology and Parmly Hearing Institute
Loyola University of Chicago, 6525 N. Sheridan Rd.,
Chicago, IL 60626 USA

INTRODUCTION

The earliest experimental interests in the hearing mechanisms of
fishes were motivated in part by questions of the functional significance of inner ear structure (e.g. Bigelow 1904; Manning 1924;
von Frisch 1938). As will be described below, the otolithic ears of
fishes do not seem suited for the type of mechanical frequency
analysis characteristic of the mammalian cochlea, and questions
arose regarding the capacities of fishes to extract information
from the acoustic waveform. Are fishes capable of frequency discrimination, and if so, how is the acoustic waveform coded by the
ear and analyzed by the brain? Without obvious mechanisms for a
frequency-domain analysis, processing in fishes is likely based on
waveform analysis in the time-domain. The fishes thus are a possible model system for studying temporal analysis which is relatively uncontaminated by a peripheral frequency analysis.

In this chapter, I review the results from a variety of experiments
on temporal auditory processing in fishes. Most of the published
data comes from my lab's work on the goldfish. Although the goldfish is not known to vocalize, and its ear is only typical of a
small group of teleosts, it is clearly well adapted to hear and
therefore must make use of sound in its daily life. In addition,
this species is easily trained in psychophysical tasks, making
possible complementary behavioral and neurophysiological experiments under the same conditions.

Overview of Fish Auditory Systems

The ears of fishes include three otolith organs; the saccule,
lagena and utricle (Fig. 1). These structures are similar to the
otolith organs of all vertebrates, and appear to be well suited for
vestibular functions; detecting linear accelerations. The otoliths
are solid calcium carbonate "stones" lying over a sensory macula
composed of supporting and hair cells. The otoliths and their
maculae vary considerably in gross and microscopic structure among
species (Popper 1977). Hair cells are organized over the macula
into two or more groups with opposing directional orientation.
Eighth nerve fibers generally innervate groups of hair cells of one
or the other directional type (a few innervate both types in the
saccule). Phase-locking in nerve fibers innervating one directional

group of hair cells essentially represents a half-wave rectified version of the stimulus waveform. However, since there are at least two opposed hair cell groups in fishes, sound waveforms are coded in greater density and detail than by the unidirectional systems characteristic of the mammals and birds.

Otolith organs function in a "vestibular" mode with the high density otolith acting as an inertial mass (Fay 1984). Acceleration of the body causes relative motion between the hair cells and the otoliths, strain across the hair cell stereocilia, and neural excitation through release of a neurotransmitter. Since body tissues are well coupled to the water environment, acoustic particle motion is an adequate stimulus for these organs. Most otolith organs respond efficiently to sound frequencies up to 200 to 300 Hz, but some are specialized for response up to 2-3 kHz. Sensitivity to particle motion has been measured to fall between 10 and 0.1 nm (Chapman and Hawkins 1973; Fay and Patricoski 1980; Fay 1984). This range is comparable to basilar membrane displacement amplitude calculated for the threshold of hearing in mammals on the basis of recent biophysical measurements (reviewed by Ashmore and Russell 1983).

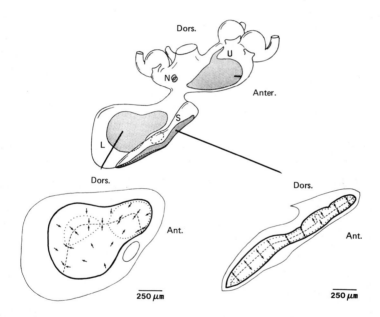

Fig. 1. Otolith organs of the goldfish; left ear, medial view, showing the maculae (stippled) of the saccule (S), utricle (U), and lagena (L). Maps of hair cell orientation patterns are shown below. The heavy line shows the macular border; dashed line shows where opposed hair cell groups abut. From Fay (1981), after Platt (1977).

The goldfish is additionally specialized to detect the sound pressure waveform through the swimbladder, which is efficiently coupled to the saccule via a set of specialized bones (Weberian ossicles). Sound pressure variations cause the swimbladder to expand and contract, and this motion is transmitted via the Weberian ossicles to the saccular endolymph where the otolith is thought to be engaged through frictional forces. The capacity to code both sound pressure and particle motion is found only among some fishes, and gives information required to determine wave impedance, direction of propagation, and acoustic intensity (Schuijf and Hawkins 1983; Fay 1984).

Overview of Fish Hearing Capacities

Fishes vary considerably in hearing sensitivity and bandwidth (Fig. 2). The most sensitive fishes (goldfish, soldierfish, and blind cavefish) have specialialized mechanical pathways for transmitting motion from the swimbladder to the ear, and respond to sound intensities of about 6×10^{-14} watt meter^{-2}. Species with intermediate sensitivity (cod, squirrelfish) have swimbladders, but no known special pathway to the ear, and those with poorest sensitivity (flatfish, tuna) lack swimbladders altogether. The 3 kHz bandwidth limitation characteristic of fishes may be an adaptation for processing only those frequencies likely to cause robust phase-locking in eighth nerve fibers.

Goldfish discriminate one pure tone frequency from another with an accuracy of about 5% for frequencies between 50 and 1000 Hz (see Fig. 16). This sensitivity puts the goldfish within the range of variation in frequency discrimination seen among the mammals and birds in this same frequency range (Fay 1973). Compared with man, however, the goldfish thresholds (and those of most mammals) are at least one order of magnitude higher. Figure 3 shows intensity discrimination thresholds for noise and 800 Hz tones measured under various conditions. Noise DLs range from about 1.2 to 2.4 dB and are independent of overall level (Weber's Law holds). This sensitivity is comparable to man's, given that listening is restricted to a 1 kHz bandwidth (Zwicker 1975). The tonal DLs obtained under conditions of continous adaptation depend strongly on overall level (Weber's Law does not hold), and thresholds may be as small as 0.13 dB, a value below that measured in any other non-human vertebrate. The tonal DLs for pulsed signals (not shown) range between 1.5 and 2.0 dB, and are independent of overall level (Fay and Coombs 1984).

This differential sensitivity places the goldfish within the range observed for most other vertebrates, with the exception of frequency discrimination compared to man. It seems that otolith organs provide the auditory system with a quality of neurally coded information rivaling that of the mammalian cochlea, for frequencies below 1 kHz or so.

While otolith organs seem unsuited for frequency analysis, single units of the goldfish's saccular nerve show some degree of frequency selectivity (Fig. 4). There is considerable variation among units in spontaneous activity (not illustrated), sensitivity (40 dB or more), best frequency (from about 100 to 800 Hz), and bandwidth. An important consequence of this peripheral filtering is the limitation of effective noise power interfering with the detection of band-limited signals. Studies of the critical masking ratio

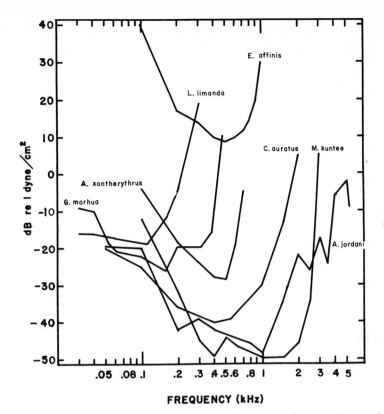

Fig. 2. Behavioral sound pressure audiograms for eight selected teleost species. Adapted from Fay and Popper (1980). G. morhua – cod; A. xantherythrus – squirrelfish; L. limanda – dab; E. affinis – tuna; C. auratus – goldfish; M. kumtee – soldierfish; A. jordani – blind cavefish.

(Buerkle 1969; Chapman and Hawkins 1973; Fay 1974; Fay and Coombs 1983), psychophysical tuning curves (Fay, Ahroon and Orawski 1978; Coombs 1981), and other masking phenomena (Tavolga 1974; Hawkins and Chapman 1975; Fay, Yost and Coombs 1983) show that the sensitivity in detecting masked tones is consistent with the existence of "auditory filters" which, in the goldfish, are most likely the tuning characteristics of saccular fibers. The origin of frequency selectivity is not known, but could possibly arise from mechanical filtering peripheral to hair cell input, stereociliar resonance, or hair cell membrane tuning. In any case, these physiological and psychophysical data complicate a simple view of the otolithic ear as "monolithic", and raise the possibility that across-fiber profiles of neural activity as well as temporal codes may underly acoustic spectral processing.

The following sections review studies designed to disentangle the roles played by frequency- and time-domain cues in hearing by the goldfish.

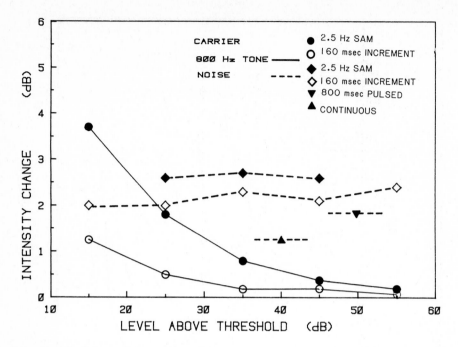

Fig. 3. Smallest detectable change in intensity for the goldfish as a function of intensity, determined using SAM (sinusoidal amplitude modulation), pulsed noise, and pulsed increments in continuous noise and tones.

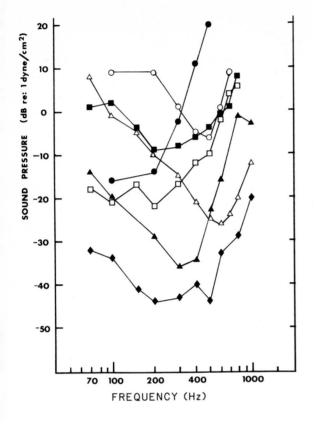

Fig. 4. Tuning curves based on synchronization critera for representative saccular units from goldfish. Adapted from Fay (1981).

TEMPORAL PROCESSING

This section considers how the auditory system of fishes encodes and processes the temporal features of sounds. "Temporal processing" is loosely defined in order to include the concepts of temporal summation, temporal resolution of envelope structure, and temporal resolution of waveform fine structure.

Temporal summation refers to the capacities of the auditory system to add up or integrate sound intensity over time, and thus to maximize the detectability of long duration signals. Two important aspects of this process are; 1) the maximum time over which intensity may be integrated (most likely a property of central mechanisms which summate neural activity), and 2) the trading relation between sound intensity and time.

Temporal resolution of envelope refers to capacities for keeping acoustic events separate in time; for coding and processing envelope features over the shortest possible time intervals so as to maximize the detection of brief signals, and to extract information from envelope patterns. This process involves at least three processes; 1) the resolution with which envelope fluctuations are represented in the periphery, 2) the sensitivity with which intensity fluctuations are coded, and 3) the times over which neural activity is sampled.

Waveform fine structure resolution refers to capacities for the neural coding and analysis of acoustic spectra in the time-domain. It is clear that acoustic waveforms are coded through a phase-locked response in peripheral neurons. However, it is less clear that a temporal code is the basis for a central spectrum analysis. In any case, the resolution with which acoustic waveforms could be analyzed in time depends on several factors including; 1) the tuning (time constants) of filters peripheral to spike generation in auditory neurons, 2) the variability with which spikes are phase-locked to the stimulus waveform, and 3) the characteristics of hypothetical central networks which, in effect, measure time intervals between spikes.

Implicit in the above definitions of temporal processing are the hypotheses that envelope integration and resolution are based on spike counting, and that waveform analysis is based on inter-spike timing. Although the following review of temporal processing provides some evidence in support of these hypotheses, alternatives have not been ruled out in many cases.

Temporal Summation

In psychophysical experiments, Offutt (1967), Popper (1972), Hawkins (1981), and Fay and Coombs (1983) studied the effect of duration on pure tone detectability. For signal durations between 10 and 500 msec, Offutt (goldfish) and Hawkins (cod) found that pure tone threshold under masking declined significantly with duration while Popper (goldfish) found no duration effect at all for detection in quiet. Fay and Coombs tested the hypothesis that the effect of duration on threshold is different in quiet and under noise masking.

Some of the results from Fay and Coombs (1983) are shown in Fig. 5 for a 400 Hz tone detected in quiet and in the presence of a broad band noise masker. Temporal summation is seen in both conditions, but the effect of duration is clearly larger for masked than for quiet detection. These functions can be thought of as duration-intensity trading relations for constant detection performance, and are well described by power functions with slopes (exponents) of -1.04 in noise, and -0.43 in quiet. Unity slope means that a doubling of duration (t) can be traded for a 3 dB reduction in threshold intensity (I), or that energy (I·t) is constant at threshold as long as t is less than 600 msec or so. In this sense, temporal summation is "perfect" for masked tone detection. In quiet, smaller changes in intensity are traded for a doubling of duration to maintain detection performance. Therefore, threshold energy grows with signal duration out to about 600 msec. The same pattern of results was obtained at different signal frequencies, at a variety of masker levels, and for noise signals as well. The exponential function of Plomp and Bouman (1959) also fit the masked threshold data quite well, estimating the maximum integration time at 590 msec. Fay and Coombs (1984) have also studied the effects of duration on the detection of noise signals added to identical

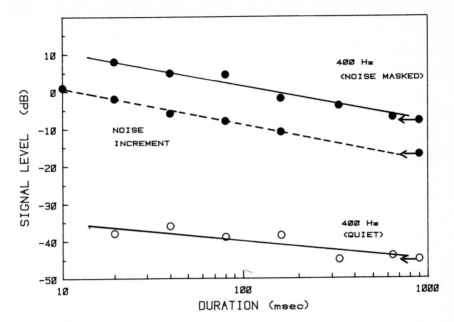

Fig. 5. Thresholds for the detection of a 400 Hz tone in quiet and in noise (-25 dB spectrum level) as a function of tone duration. Also thresholds for detecting an increment in noise level (in-phase addition). For tone detection, ordinate is signal level in dB re: 1 dyne/sq cm. For increment detection, ordinate is S/N in dB.

(correlated) continuous noises. Here again, signal level is a power function of duration with an exponent of -0.96, and an estimated time constant of 450 msec (dasked line in Fig. 5).

Zwislocki (1969) outlined a simple model for temporal summation which rests on the idea that the central integrator operates not on acoustic intensity, but on neural activity input from the periphery. If detection decisions are made on the basis of the attainment of a given spike count within the integration time, then the trading relation between sound intensity and duration (the slope of the temporal summation function) will be determined, at least in part, by the relation between spike rate and stimulus intensity (the slopes of rate-intensity functions). Specifically, Fay and Coombs (1983) interpreted Zwislocki's (1969) analysis as follows:

1) for constant energy at threshold, signal intensity (I) times duration (t) is a constant
2) but more realistically, that spike rate (R) times duration (t) equals a criterion spike count (C).
3) since R is proportional to signal intensity (I) raised to a power (n) (see below),
4) then I^n times t equals C
5) and threshold intensity (I) is proportional to $t^{(-1/n)}$.

In other words, the exponents of psychophysical temporal summation functions should be equal to the negative reciprocal of rate-intensity power function exponents for auditory neurons. These rate-intensity functions were measured for tone and noise signals, in quiet and under masking, in goldfish saccular nerve fibers (Fig. 6). Near threshold (and at low spike rates), these relations are very well described by power functions with exponents that range between 2 and 4 in quiet; values considerably greater than those observed in the normal mammalian eighth nerve. Masking noise reduces the exponents of rate-intensity functions to values equal to or less than 1. These values account well for the psychophysical temporal summation function exponents of -0.4 and -1.0 obtained in quiet and in noise, respectively.

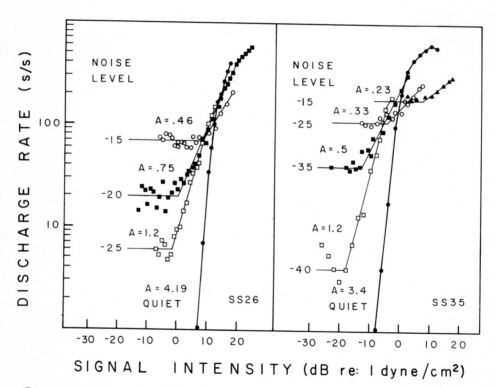

Fig. 6. Spike rate-intensity functions for two non-spontaneous saccular neurons in quiet and in the presence of a continuous noise masker at spectrum levels indicated. The exponents (A) of the power functions shown (lines) are indicated. From Fay and Coombs (1983).

These results suggest that central detection mechanisms exist in fishes which integrate neural activity with a time constant in the region of 450-600 msec, and that the intensity-duration trading ratio is determined by the slopes of rate-intensity functions for peripheral fibers. These results for the goldfish differ from those of mammals (including man) and birds (Dooling 1980) in that the integration time is longer in goldfish (450-600 msec versus about

200 msec), and temporal integration for the goldfish in quiet shows shallower slopes (-0.4 versus approximately -1.0). The latter difference is consistent with the very steep rate-intensity functions for some goldfish eighth nerve units compared with those of mammals (Zwislocki 1973). Assuming that similar central summation mechanisms also operate in mammals and birds, this analysis predicts that manipulations which cause steeper rate-intensity functions could result in shallower temporal summation functions. Evidence for this comes from mammalian work showing that trauma to the ear can cause both steepening of rate-intensity functions in cochlear neurons (Schmiedt and Zwislocki 1980), and abnormally shallow temporal summation functions (Jerger 1955; Henderson 1969).

Temporal Envelope Resolution

The sensitivity and temporal resolution for detecting envelope fluctuation has been studied in fishes using several different paradigms. These include the minimum detectable temporal gap in a continuous sound (Fay and Coombs 1984), the sensitivity for detecting sinusoidal amplitude modulation (SAM) as a function of modulation rate (the temporal modulation transfer function or TMTF) (Fay 1980; Coombs and Fay 1984), the smallest detectable change in the repetition rate for sound bursts (Fay 1982; Fay and Passow 1982), and the detectability of brief signals which just follow maskers in time (the time course of forward masking) (Popper and Clarke 1978). While each of these paradigms measures some aspect of temporal resolution, they do not estimate a single global characteristic such as "minimum integration time" (Green 1973), in part because the underlying neural codes and analyzing mechanisms may be different, as illustrated below.

Gap detection experiments measure the duration of the shortest silent interval in noise which is just detectable. Using wide band noise, gaps in the range of 2 to 3 msec are just detectable in man (Fitzgibbons 1984; Buus and Florentine, this volume), and in birds (Dooling et al. 1978). This measure of the minimum integration time is ordinarily thought to reflect the properties of a central system which measures a "running" average spike rate with some averaging time constant. Recent studies from our lab on gap detection in the goldfish measured 1) the minimum gap "depth" (intensity decrement) as a function of gap duration, and 2) the smallest detectable complete gap (silent interval). Fig. 7 shows these data for both continuous noise and tone signals.

For noise, the threshold for intensity decrements is about 3.3 dB for 160 msec "gaps", rises rapidly toward the shorter durations, and disappears completely for gaps less than about 35 msec. Thus, the capacity to detect "gaps" depends on both the depth of the gap and on its duration, and temporal acuity for the goldfish by this measure is significantly inferior to that for birds and mammals. It should be pointed out, however, that gap detection in man depends on sound pressure level, and on the center frequency and bandwidth of the noise (Fitzgibbons 1984). When human observers listen in the goldfish's bandwidth of hearing (low-pass at about 1 kHz), the minimum gap is about 10 msec (see also Buus and Florentine, this volume).

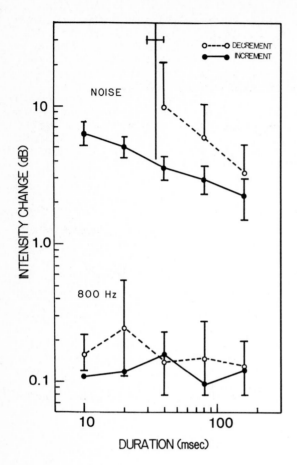

Fig. 7. Behavioral thresholds (means and standard deviations) for the smallest detectable change in intensity (increments and decrements) of continuous tone and noise signals at about 40 dB above threshold. The vertical line with error bars at 35 msec is the minimum detectable gap. From Fay and Coombs (1984).

Goldfish saccular units code all gaps in noise with an immediate cessation of firing, as illustrated in the PST histograms of Fig. 8. It thus appears that the relatively poor psychophysical performance of the fish cannot be simply explained on the basis of an inadequate neural representation of silent intervals. However, one interesting feature of the neural response is the transient burst of spike activity which follows the gap (see Evans, this volume). This burst depends on the depth of the gap and its duration in much the same way that detectability depends on these two variables. It is possible, then, that gap detection in the goldfish is based on the increase in spike rate following the gap. The implication that increases in spike rate are more detectable than decreases is consistent with the findings of Sinnott (1984) that animals are particularly poor at discriminating downward shifts in stimulus intensity compared with human adults. If gap detection by the goldfish is based on the detection of spike rate increases, then it seems that the gap thresholds reflect more the time course of recovery from adaptation than the time constant of a central integrator.

Increment and decrement thresholds for a tone (Fig. 7) are remark-

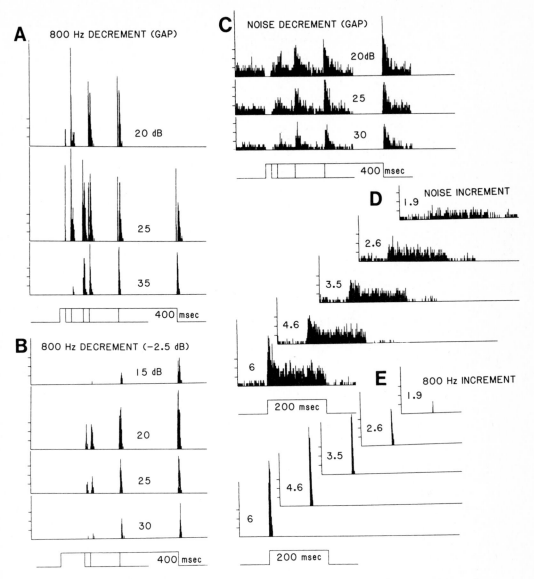

Fig. 8. PSTs for saccular fiber in response to increments and decrements in continuous sounds. A. Responses following complete decrements (gaps) in a tone at different gap durations (20, 40, 80, 100, 200, 400 msec) at 3 levels of overall attenuation. The highest sound level (atten. = 20 dB) is +32 dB re: 1 dyne cm^{-2}. PSTs for different durations are superimposed. B. Same as A with decrement depth of -2.5 dB. C. Same as in A with a noise carrier. D. Different noise increments (dB). E. Same as D for a tone. Tic marks on the ordinate show instantaneous spike rates of 100, 200 and 300 spikes sec^{-1} (1 msec bin widths).

able in that the sensitivity for detecting both decrements and increments is quite high (0.1 to 0.2 dB) and there is no clear effect of duration between 10 and 160 msec. These effects could be due to "off-frequency" listening since the onset of the increment is accompanied by a "splatter" of energy to different frequency regions where the effective signal-to-noise ratio may be more favorable. In human studies, Leshowitz and Wightman (1971) found that for abrupt increment transitions (and at high sound levels) where splatter is maximal, increment thresholds were 0.1 dB and independent of duration. When band pass filtering was used to attenuate spectral spread, thresholds increased and showed a duration effect which was nearly that predicted assuming perfect temporal integration. Although we attempted to reduce splatter using a 3 msec rise-fall time in addition to a band pass filter centered at the tone frequency, it is possible that our results can be explained similarly.

However, evidence suggests that other processes may underly this sensitivity which are different from those apparantly operating in human hearing. First, the goldfish auditory system is relatively poorly designed for the kind of frequency analysis required to make use of spectral splatter as a cue for detection. Second, increment thresholds were determined at about 40 dB above absolute detection threshold; a level at which the effects of splatter are small. Third, this same sensitivity for amplitude fluctuation was measured when animals detected sinusoidal amplitude modulation (SAM) of an 800 Hz tone, significantly exceeding that measured for man under identical conditions (Fay 1980).

This latter point is illustrated in Fig. 9 showing behavioral and neural temporal modulation transfer functions (TMTF) for noise and tones. The behavioral thresholds for tones show a band pass characteristic with a best sensitivity (peak-trough intensity difference of 0.1 dB) at about 200 Hz, and a 3 dB per octave high pass slope. This contrasts sharply with the function for man which shows at first low pass behavior (presumably reflecting an integration mechanism which limits temporal envelope resolution), and then a rise in sensitivity which reflects the detection of side bands through frequency-domain spectral analysis. The goldfish tonal TMTF cannot be accounted for similarly, either by an integration process at low modulation rates, or by frequency-domain processing at high rates. Correspondence is also lacking between human and goldfish data for the noise TMTF. The human function shows low pass behavior to be expected from an integration mechanism. The goldfish function, on the other hand, is flat out to 400 Hz with an overall lower sensitivity. Taken at face value, this function suggests an integration time which is less than 0.4 msec. Alternatively, it is possible that this method of estimating the minimum integration time is inappropriate for the goldfish due to non-linearities in peripheral envelope coding.

TMTFs for saccular fibers (Fig. 9) were obtained similarly by determining the modulation depth producing a criterion degree of synchronization to the modulation envelope at different modulation rates. Two points are noteworthy: fibers differ with respect to sensitivity and best modulation rate, and the ensemble sensitivity fairly well parallels the psychophysical functions for both noise

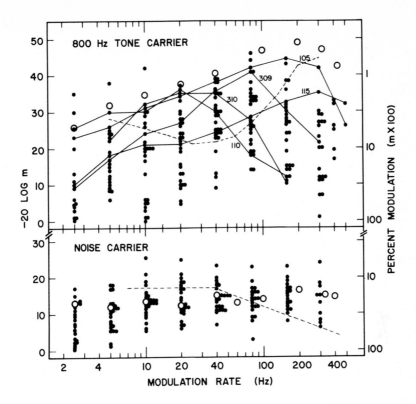

Fig. 9. Temporal modulation transfer functions (TMTF) for human observers (dashed line), goldfish observers (large open circles), and for saccular units (dots). The data for selected units are connected with lines. Adapted from Fay (1980).

and tones. Clear non-linearities are evident in these data, and in the results of Fig. 10 showing envelope synchronization in a representative saccular fiber. First, fibers show a considerable "gain" in the sense that spike rate modulation far exceeds that of the stimulus envelope. For example, spike rate is modulated by nearly 100% in response to envelope fluctuations of about 3%. Second, fibers show a form of "temporal tuning" to SAM tones with best modulation rates ranging from 20 to 160 Hz. These effects are consistent with the psychophysical data (Fay 1980) but clearly different from those observed in mammals (e.g. Frisina et al. 1984) (see also Capranica, this volume). What processes underly these effects?

Using both intra- and extra-cellular recording from saccular fibers, Furukawa and Matsuura (1978), Furukawa, Hayashida and Matsuura (1978), Furukawa, Kuno and Matsuura (1982), and Kuno (1983) have analyzed in detail the nature and causes of adaptation in the goldfish. They demonstrated, first, that the adaptation of firing is a synaptic effect due to the reduction in epsp amplitude following tone onset. Adaptive changes in epsp amplitude (and spike probability) are due to changes in the number of hypothetical "ac-

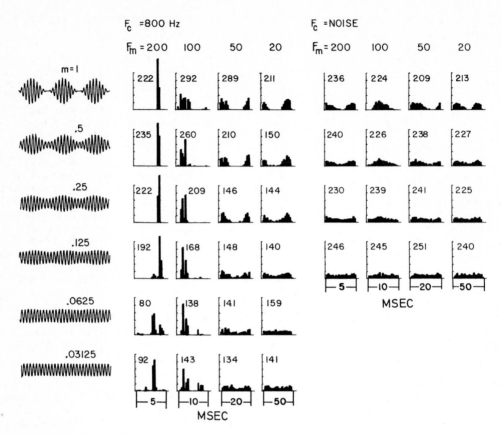

Fig. 10. Modulation period histograms for a saccular fiber in response to SAM tone and noise signals at different modulation depths (m) and rates (F_m). The numbers indicate average spike rate. From Fay (1982).

tive release sites" for neurotransmitter, and not to changes in the probability of quantal release. This leads to the notion that release sites differ widely in "threshold" so that, for example, a given adapting tone depletes all sites with low thresholds, but leaves higher threshold sites completely intact and available for response to intensity increments. In addition, Furukawa and associates have observed that in response to small (1-2 dB) decrements in intensity, epsps may transiently disappear altogether. This led to the hypothesis that the replenishment of neurotransmitter occurs serially, from high threshold sites to low. In this case, for example, the adapted epsps were produced by release of transmitter from the only sites being replenished (the highest threshold sites activated by the adapting sound). At this point, a reduction in intensity brings the stimulus below the threshold for any replenished site, and the epsp disappears for the time it takes the transmitter to trickle down to lower threshold sites where release can take place.

Our observations of extra-cellular spike activity in response to intensity increments are at least qualitatively consistent with aspects of this synaptic model (Fig. 8). Two features of the response are noted. First, the unit shows a greater sensitivity to tone than to noise decrements in the "recovery" burst at decrement offset. Second, the recovery occurring during the decrement depends on duration and depth of the decrement, with longer recovery times occurring for the smaller decrements (compare Fig. 8A and 8B). This appears not to be in accord with the notion that transmitter replacement occurs serially, from high threshold sites to low. The recovery time varies among neurons, and this may account for the variation in "temporal tuning" shown in Fig. 9. For example, the response of a unit to an SAM tone initially grows with modulation rate, but then begins to decline, presumably when the modulation period becomes short relative to the unit's recovery time. In combination with the very steep rate-intensity functions charac-teristic of some saccular units, these synaptic effects may be the basis for 1) the extreme sensitivity of the goldfish in detecting increments, decrements, and SAM under continuous adaptation, 2) the variation in temporal tuning seen among saccular fibers, 3) and the shape of the tonal TMTF measured behaviorally. The apparent lack of temporal summation in detecting tone increments and decrements (Fig. 7) is correlated with the very phasic (rapidly adapting) response to increases in intensity seen in continuously adapted saccular units. In other words, the number of spikes evoked phasically at the onset of a signal is independent of the signal's duration.

Most aspects of the behavioral and physiological response to noise signals is quite different from those evoked by tones, particularly under continuous adaptation. Adaptation to tones is often complete (spike rate goes to zero), while much less adaptation occurs to noise. The sensitivity in detecting increments and decrements in noise signals is considerably less than for tones. The behavioral and neural TMTFs for noise are much less sensitive and lack the dependence on modulation rate compared with tones. Temporal summa-tion is always observed in detecting noise signals (or tone signals in a noise background), but may be absent altogether for tones in quiet and under continuous tonal adaptation. Finally, Weber's Law holds for noise increment detection, but fails significantly for tones in that the increment threshold (in dB) declines precipi-tously toward high sound pressure levels (Hall, Patricoski and Fay 1981; Fay and Coombs 1984). In the light of Furukawa's conceptions of the hair cell-nerve fiber synapse in the goldfish saccule, these differences could be accounted for by essential dissimilarities in the instantaneous amplitude distributions for these two signal types. Those for tones are deterministic; the peak amplitude always reaches and never exceeds a given value. For long term tone adaptation, transmitter release sites fall into just two categor-ies; those whose thresholds fall below the peak tone level (and are depleted), and those with higher thresholds (totally intact). For noise, the amplitude distribution is stochastic; amplitudes vary randomly according to a Gaussian distribution. In this case, a site's threshold determines its probability of being replenished (active) at any given instant, and thus the number of active sites varies continuously over the threshold continuum. This may explain why adaptation is never as complete for noise stimulation, and why the goldfish auditory system behaves more "linearly" under noise masking. Noisy signals are also subject to temporal summation because the neural response is more tonic (particularly near threshold), and spikes accumulate nearly linearly with time.

In summary, it appears that while we are beginning to understand the coding of temporal envelope structure and its relations to psychophysical data on envelope detection, both the gap detection and the TMTF paradigm have apparently failed to give a valid estimate of the goldfish's minimum integration time. Based on mammalian data, this limit on resolution is ordinarily thought to be a characteristic of central neural systems which compute something like a running spike rate averaged over some short integration time (2-3 msec). Psychophysical results on envelope resolution for the goldfish seem to be dominated by adaptation and recovery processes at the peripheral synapse, by steep rate-intensity functions in primary fibers, and by the apparent inability of the auditory system to use transient decreases in spike rate as useful information.

Waveform Processing

Given that saccular fibers show some frequency selectivity, the ability to detect small (5%) increments in tone frequency could be based on either a change in the profile of activity across an array of fibers, or on a change in the inter-spike-interval distributions within fibers. Fay (1978) investigated the adequacy of the temporal coding hypothesis by measuring the accuracy with which saccular fibers phase-lock to tones (at the levels used in the behavioral frequency DL measurements). Figure 11 shows the relation between the behavioral frequency DLs and phase-locking accuracy. There are two major points here. First, both the period DL and phase-locking accuracy increase with frequency at the same rate, and the error in msec is approximately a constant proportion of the tone period. Second, the smallest neural errors at any given frequency are about equal to the behavioral errors (period DLs).

A simple temporal hypothesis for frequency discrimination is that the stimulus period is estimated through measurements of the times between spikes. The discrimination problem could be a decision as to whether two samples of inter-spike-intervals are drawn from the same or different populations. Assuming that the means of these populations are equal to the periods of the two signals to be discriminated, and that the variance is completely determined by phase-locking error, then threshold-like behavior should be expected when the difference between the means (the period DL) is equal to the distribution's standard deviation (phase-locking error). This is the case for units showing the least phase-locking error. This is correlative evidence that frequency discrimination is a matter of discriminating one distribution of neural time intervals from another. At 1000 Hz, the resolution, or "accuracy" (see Menne, this volume), of the goldfish auditory system seems to be about 40-60 μsec.

Fay and Passow (1982) attempted to measure the temporal resolution of the goldfish auditory system more directly. Animals detected a change in the period at which short sound bursts were repeated,

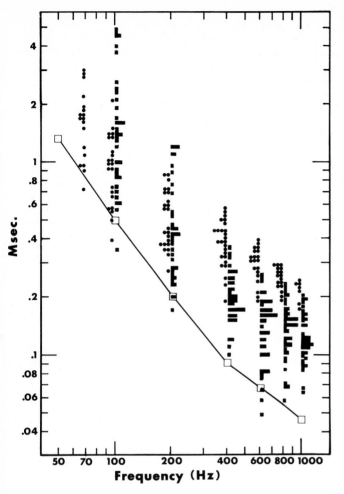

Fig. 11. The line con-
necting the squares shows
behavioral pure tone period
discrimination thresholds
in msec (derived from frequency DLs as shown in Fig.
16), compared with phase-
locking error in saccular
units (standard deviation
of period histograms, in
msec) as a function of tone
frequency. From Fay (1978).

under two conditions. In one condition, the bursts were perfectly
periodic, repeated at intervals of 5 or 10 msec. In the other condition, a random fluctuation was introduced in each successive
inter burst interval with a constant mean interval. This random
temporal "jitter" in stimulus intervals was assumed to cause an
identical jitter in the neural code which would simply add to the
internal temporal jitter (the quantity we wished to measure). Animals detected a slow sinusoidal change in mean inter burst interval
(ΔP) for different amounts of "stimulus-domain" jitter introduced
(Fig. 12). The lowest levels of stimulus jitter had no effect on
the discrimination, but at the highest levels of jitter, the ΔP
tends to equal the stimulus jitter (points on the straight line
with unity slope). The two curves fit to the data describe the
relation $\Delta P=(S_e^2 + S_i^2)^{0.5}$, or the independent addition of two
random processes. The parameter S_e defines the standard deviation
of external jitter, and S_i can be identified as an hypothetical

internal temporal jitter independent of S_e. The curves were fit by finding the value of S_i which minimized the difference between the data and the model function. These values are 160 μsec for the 200 Hz repetition rate (5 msec period), and 710 μsec for the 100 Hz rate (10 msec period).

The squares referred to the ordinate are the ΔP thresholds determined without external jitter (S_e=0), and were not used in fitting the function. These independent estimates of internal jitter correspond closely to those derived from the curve fitting procedure, and provide further evidence that this discrimination is limited by the internal representation of inter burst interval in the time-domain. On the abscissa, the squares are the thresholds for the detection of RMS random jitter imposed on the otherwise periodic burst signals. These values (about 60 and 240 microsec for the 5 and 10 msec burst periods, respectively) are a factor of about $2\sqrt{2}$ less than the peak-peak ΔP thresholds determined for sinusoidal modulation of inter burst interval (with S_e=0), indicating that the RMS ΔP values are the same whether jitter is introduced sinusoidally or randomly.

In summary, these experiments suggest that burst interval discrimination thresholds are determined by the variability with which the intervals are represented in a temporal neural code. These results are in accord with data from the pure tone frequency discrimination studies (described above), and support the notion that the goldfish solves both types of problems (waveform fine structure and envelope periodicity discrimination) through time-domain processing. In this context, the temporal resolution of the goldfish auditory system is frequency-dependent, ranging from 700 μsec at 100 Hz, to 50 μsec at 1 kHz.

What are the Roles of Frequency-Domain Cues

The foregoing experiments demonstrate the adequacy of a temporal coding hypothesis in accounting for the psychophysics, but do not rule out the possibility that the problem is solved in the frequency-domain (i.e. through processing the profile of activity across arrays of tuned peripheral channels). Fay and Passow (1982), and Fay (1982) obtained discrimination thresholds for sound burst interval using a variety of signals with different degrees of temporal and spectral structure (Fig. 13). The sinusoidally amplitude modulated (SAM) tone and the periodic filtered clicks have line spectra, so that any change in periodicity produces both frequency and time cues. The SAM noise, and the two types of gated noise have continuous long term spectra which are not affected by changes in repetition period, and therefore lack frequency cues. The difference between types I and II gated noise is in the degree to which the envelope is defined over the short term. Note that the burst-to-burst fluctuation in amplitude for type II noise is considerably greater than that for type I. (See Fay and Passow 1982 for details of signal definition).

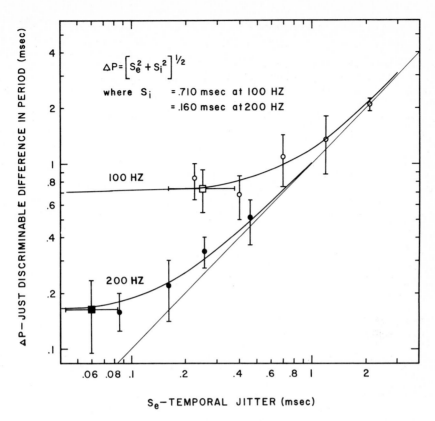

$$\Delta P = \left[S_e^2 + S_i^2 \right]^{1/2}$$

where S_i = .710 msec at 100 HZ
= .160 msec at 200 HZ

100 HZ

200 HZ

S_e—TEMPORAL JITTER (msec)

ΔP—JUST DISCRIMINABLE DIFFERENCE IN PERIOD (msec)

Fig. 12. Behavioral thresholds for the detection of changes in sound burst rate as a function of random jitter in burst rate. From Fay and Passow (1982).

Period discrimination thresholds (ΔP) for each signal type are shown in Figure 14. The three nearly coincident lower curves are for the SAM tone and filtered clicks (both of which contain time and spectral structure), and type I gated noise (which is devoid of long term spectral structure). Since the addition of spectral structure has no effect on the period discrimination thresholds, we conclude that the information it provides is less useful to the goldfish than the temporal structure.

The remaining two curves are for SAM and type II gated noise, both of which lack spectral structure but vary in bandwidth and the degree to which the envelope is defined over the short term. A comparison with the type I noise thresholds shows that envelope definition determines how well changes in envelope periodicity are discriminated, with the lowest thresholds occurring for the best defined envelopes. The same conclusion was reached from experiments on the effect of modulation depth on modulation period discrimination using SAM tones (Fay 1982). In summary, we have not been able to show any effects of spectral structure on the ability to discriminate changes in envelope periodicity. At the same time, all

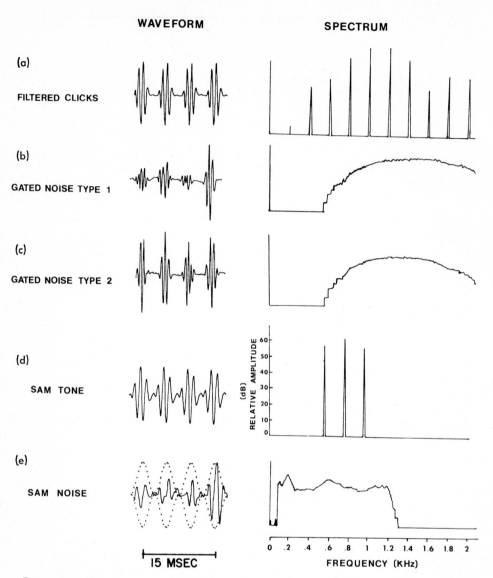

Fig. 13. Waveforms and spectra for various signals used in sound burst rate discrimination experiments. From Fay and Passow (1982).

of the psychophysical data reviewed are consistent with the hypothesis that these envelope discriminations are based on time-domain processing.

Repetition Noise Processing

Questions of the extent to which man uses time and frequency cues to analyze auditory signals most often arises in the context of

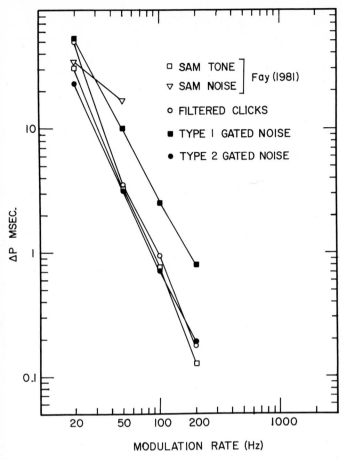

Fig. 14. Behavioral thresholds for the detection of sound burst rate changes for the signals shown in Fig. 13. From Fay and Passow (1982).

experiments on "periodicity pitch". It has long been recognized that the time and frequency structures of complex periodic signals are two complementary dimensions along which analysis may take place. Since the human auditory system is clearly capable of analyzing within both dimensions, the relative roles played by time and frequency in evoking periodicity pitch perceptions has been difficult to determine, and many questions remain open today (see Langner, this volume). The following experiment used psychophysical and neurophysiological methods to study how the otolithic ear of the goldfish codes these dimensions in repetition noise (a particular waveform type often used in human pitch experiments), and to what extent these neural representations may be used in psychophysical behavior.

Repetition noise is created in the simplist case by adding a wide band noise to a delayed (and possibly attenuated) replica of itself (see Fig. 15). Such signals may occur in the natural world when sound reaches the ear directly and via a single reflective path. When such echo delays are in the range of 0.5 to 50 msec, human observers may describe a sensation of "coloration" which may be

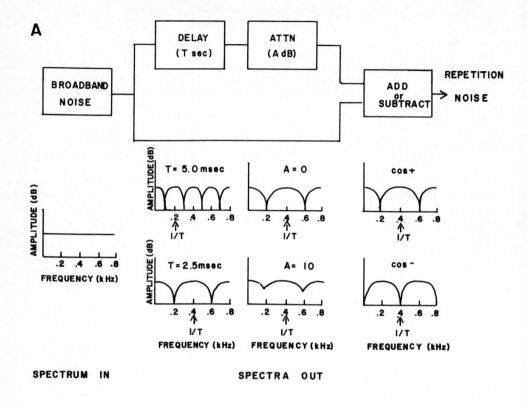

A

SPECTRUM IN   **SPECTRA OUT**

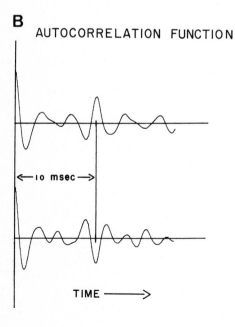

B

AUTOCORRELATION FUNCTION

←10 msec→

TIME ——→

Fig. 15. A. Schematic block diagram showing how repetition noise is generated, and the effects as the OUTPUT spectrum of varying delay (T) and attenuation (A). B. Autocorrelation functions for COS+ noise with delay = 10 msec (TOP), and for COS- noise with delay = 10 msec. From Fay, Yost, and Coombs (1983).

defined as pitch. Generally, the pitch value in Hz is equal to the reciprocal of the delay (t) in seconds. The spectra of such signals have a sinusoidal profile with the first power peak occurring at 1/t Hz, and successive peaks at integral multiples of this frequency. I refer to this signal as cos+ noise. If the delayed echo is inverted before summing with the direct signal, repetition noise is created having a first power minimum at 1/t Hz, and successive minima at integral multiple frequencies. This signal, referred to as cos- noise, seems to have two pitches, one just above and one just below 1/t Hz. Although a casual look at the waveform of repetition noise reveals no obvious time structure, the autocorrelation functions show a prominence located in the region of t sec.

For human observers, a change in the delay Δt used in producing repetition noise may be detected if the values of t and Δt fall in the proper ranges (T = 2-50 msec, Δt = 0.03 t), and a controversy exists today to what extent this detection is based on frequency- or time-domain processing of the signal. Fay, Yost, and Coombs (1983) asked goldfish to detect changes in the delay of repetition noise under a variety of conditions, and compared psychophysical performance with the ways in which the delay values are coded in the discharge patterns of saccular units.

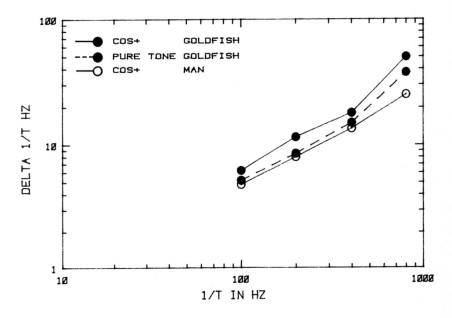

Fig. 16. Behavioral thresholds for the detection of changes in repetition noise delay in goldfish and man, and pure tone frequency discrimination thresholds for goldfish. From Fay, Yost and Coombs (1983).

Figure 16 shows the smallest detectable change in delay as a function of delay for cos+ noise (cos- thresholds did not differ significantly), for the goldfish and for man as determined by Yost, Hill, and Perez-Falcon (1978). Also included in the figure are the goldfish's pure tone discrimination thresholds (Fay 1970). The essential point is that changes of about 6-7% are detectable at delays between 1.25 and 10 msec. The function for cos+ noise is quite similar to that for pure tone frequency discrimination in the goldfish, and both of these parallel the human cos+ noise discrimination function. (It should be pointed out that Bilsen and Weiman (1980) and Johnson (1980) have determined delay discrimination thresholds for man to be about 1% using somewhat different psychophysical and stimulus generation methods). In any case, it is noteworthy that goldfish discriminate cos+ noise "pitch" changes with about the same sensitivity as they discriminate pure tone frequency, and the difference between goldfish and man in cos+ noise discrimination is far less than their difference in pure tone frequency discrimination. This suggests that the goldfish codes and analyzes cos+ noise delay and pure tone frequency using the same mechanism, and that this may be essentially the same as used by man in cos+ noise discrimination.

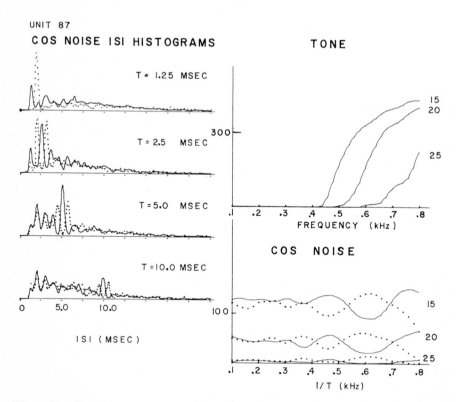

Fig. 17. Response of a goldfish saccular unit to repetition noise (and pure tones) as a function of noise delay (and pure tone frequency). From Fay, Yost, and Coombs (1983).

Discharge patterns from saccular units were recorded in response to cos+ and cos- noise, and to pure tones in order to determine how cos noise delay is coded in peripheral fibers. Figure 17 shows the results of three experiments performed on one representative fiber. In the upper right, average spike rate is plotted as s function of pure tone frequency at several overall levels. It is clear that this unit is broadly tuned in the region of 700-800 Hz. At bottom right, average rate is plotted as a function of cos+ and cos- noise delay at two overall levels. Although spike rate depends only weakly on delay, the functions tend to show an oscillation which is out of phase for the cos+ and cos- signals, and which declines toward the longer delays (lower values of 1/t Hz). This is the pattern to be expected for a unit broadly tuned in the 700-800 Hz region. As 1/t decreases from 800 Hz, the response to cos+ noise first declines, reflecting the movement of the first spectral peak toward lower frequency regions, and out of the tuning curve. The response then recovers toward 400 Hz as the second spectral peak moves into the tuning curve. The cos- noise produces the opposite pattern since it's peaks correspond to the minima of the cos+ spectrum. The effect of delay on spike rate declines toward the lower 1/t values because the spacing between spectral peaks becomes small relative to the tuning curve width.

The left panel shows inter-spike-interval distributions for the response evoke by continuous cos noise. In this cell, as in every one studied, the cos+ ISI distribution tends to have a peak at the interval corresponding to the delay, and the cos- ISI distribution tends to have a null at the interval of the delay with minor flanking peaks. In this respect, the ISI distributions resemble the autocorrelation functions for cos noise. For short delays, the limitation to this is the refractory period (in the region of 1 msec). In one sense, there is no limit to the size of the longer delays that can be represented as a peak in the ISI distribution. However, given that the unit responds even moderately to the noise signal, the probability of ISIs corresponding to long delays declines rapidly with the magnitude of the delay. These two factors would seem to set upper and lower boundries on the ISI code for delay.

These data show that the otolithic ear codes a change in cos noise delay in two ways: In changes in spike rate occurring within and between saccular units, and in changes in the location of ISI distribution peaks within units. There are several aspects of the data which suggest the latter temporal code as more likely. First, the greatest variation in spike rate with delay occurs at the shortest delay values (1.25-2.5 msec), and show essentially no variation in the region of 10 msec. Yet, delay discrimination thresholds are lowest for the longest delays. Second, a comparison of the spike rate changes produced by shifting pure tone frequency with those produced by shifting cos noise delay shows that pure tone frequency has a far larger effect. Yet, discrimination thresholds for tone frequency are nearly identical to those for cos noise delay, as shown in Figure 15. These and other considerations (see the "pitch strength" experiment in Fay, Yost, and Coombs 1983) lead to the conclusion that both pure tone frequency and cos noise delay are discriminated on the basis of information contained in inter-spike-interval distributions, and not upon profiles of activity across arrays of saccular fibers. We have argued that applying this temporal coding hypothesis for cos noise delay to the human psychophysical data leads to possible explanations for the "existence region" for cos noise pitch, the "dominance region" phenomenon, and for the pitch ambiguities associated with cos noise.

SUMMARY AND CONCLUSIONS

Studies on the psychophysics and neurophysiology of hearing in the goldfish have focused on the coding of temporal patterns by a simple otolithic ear, and on the processing of these codes by the brain. Temporal analysis is ultimately limited by the low frequency bandwidth of hearing typical of fishes, but good use of this bandwidth for time-domain processing is reflected in rather broadly tuned peripheral channels. All sounds which fishes hear cause phase-locking in peripheral fibers, and the bi-directional arrangement of hair cells enhances the density and detail of phase-locked information compared with mammals and birds. Given an adequate peripheral representation of a sound's duration, fishes show temporal summation at threshold (time constant of 450-600 msec); a process probably typical of all vertebrate sensory systems. Temporal resolution of envelope patterns has been difficult to relate to similar data for mammals due to extraordinarily steep rate-intensity functions, variation in patterns of adaptation in saccular fibers, and other non-linear phenomena of the hair cell-nerve fiber synapse. Gap detection acuity for noise (35 msec) is limited by the goldfish's low frequency hearing range, and possibly by an inability to process short term rate reductions in auditory nerve fibers. Fishes seem to process inter-spike-interval information to analyze waveform with a temporal resolution of 4-6% (50-1000 μsec depending on frequency between 1000 and 100 Hz). Fishes seem to be particularly well adapted for temporal processing, and poorly adapted for frequency-domain processing. No behavioral evidence has been found for spectrum analysis based on across-fiber profiles of neural activity. The adaptive significance of time-domain processing is in extracting information from vocalization sounds (Myrberg and Spires 1972), in locating the position of sound sources (Schuijf 1981; Schuijf and Hawkins 1983), and in acquiring general information on the structure of the acoustic environment (Fay, Yost, and Coombs 1983; Fay 1984).

ACKNOWLEDGEMENTS

This research was supported by research grants from the NSF (BNS-8111354) and the NIH (NS-15268) and a Research Career Development Award from NIH (NINCDS) to R. Fay. Much of this work was carried out with collaborators, and the contribution of Dr. Sheryl Coombs is particularly acknowledged. Thanks are also due to Bill Yost and Toby Dye for much helpful discussion. Thanks also to Axel Michelsen for flawless organizational work on the Symposium and for editorial suggestions. Thanks to the Danavox Foundation for their support for this presentation.

REFERENCES

Ashmore JF, Russell IJ (1983) The physiology of hair cells, in Bioacoustics: A Comparative Approach (B Lewis, Ed.), Academic Press,
London, p 149-180

Bigelow HB (1904) The sense of hearing in the goldfish. Amer Natura-
list 38:275-284

Bilsen F, Weiman J (1980) Atonal periodicity sensation for comb
filtered noise signals, in Psychophysical, Physiological and
Behavioral Studies in Hearing (G van den Brink, F Bilsen, Eds.).
Delft University Press, p 379

Buerkle U (1969) Auditory masking and the critical band in Atlantic
Cod (Gadus morhua). J Fish Res Board Can 26:1113-1119

Buus S, Florentine M (1984) This volume

Capranica R, Rose GJ, Brenowitz EA (1984) This volume

Chapman CJ, Hawkins AD (1973) A field study of hearing in the cod
(Gadus morhua). J Comp Physiol 85:147-167

Coombs SL (1981) Interspecific differences in hearing capabilities
for select teleost species, in Hearing and Sound Communication in
Fishes (W Tavolga, A Popper, R Fay, Eds.). Springer-Verlag, New
York, p 173

Coombs SL, Fay RR (1984) Factors affecting detection of amplitude
fluctuations by the goldfish auditory system. Neuroscience Ab-
stracts

Dooling R (1980) Behavior and psychophysics of hearing in birds, in
Comparative Studies of Hearing in Vertebrates (A Popper, R Fay,
Eds.). Springer-Verlag, New York, pp 261-288

Dooling RJ, Zoloth SR, Baylis JR (1978) Auditory sensitivity, equal
loudness, temporal resolving power and vocalizations in the house
finch (Carpodacus mexicanus). J Comp Physiol Psychol 92:867-876

Evans E (1984) This volume

Fay RR (1970) Auditory frequency discrimination in the goldfish
(Carassius auratus). J Comp Physiol Psychol 73:175-180

Fay RR (1973) Auditory frequency discrimination in vertebrates. J
Acoust Soc Am 56:206-209

Fay RR (1974) Masking of tones by noise for the goldfish (Carassius
auratus). J Comp Physiol Psychol 87:708-716

Fay RR (1978) Coding of information in single auditory-nerve fibers
of the goldfish. J Acoust Soc Am 63:136-146

Fay RR (1980) Psychophysics and neurophysiology of temporal factors
in hearing by the goldfish: Amplitude modulation detection. J
Neurophysiol 44:312-332

Fay RR (1981) Coding of acoustic information in the 8th nerve, in
Hearing and Sound Communication in Fishes (W Tavolga, A Popper, R
Fay, Eds.) Springer-Verlag, New York, p 189

Fay RR (1982) Neural mechanisms of auditory temporal discrimination
by the goldfish. J Comp Physiol 147:201-216

Fay RR (1984) The goldfish ear codes the axis of acoustic particle
motion in three dimensions. Science 225:951-954

Fay RR, Ahroon W, Orawaski A (1978) Auditory masking patterns in
the goldfish (Carassius auratus): Psychophysical tuning curves. J
Exp Biol 74:83-100

Fay RR, Coombs S (1983) Neural mechanisms of sound detection and
temporal summation. Hearing Research 10:69-92

Fay RR, Coombs SL (1984) Sound intensity processing by the goldfish
auditory system. J Acoust Soc Amer, 76:S 12

Fay RR, Passow B (1982) Temporal discrimination in the goldfish. J Acoust Soc Amer 72:753-760

Fay RR, Patricoski M (1980) Sensory mechanisms for low frequency vibration detection in fishes, in Abnormal Animal Behavior Preceding Earthquakes: Conference II (R Buskirk, Ed.). U.S. Geological Survey Open File Report, pp 80-453

Fay RR, Popper A (1980) Structure and function of teleost auditory systems, in Comparative Studies of Hearing in Vertebrates (A Popper, R Fay, Eds.). Springer-Verlag, New York, p 1

Fay RR, Yost WA, Coombs SL (1983) Psychophysics and neurophysiology of repetition noise processing in a vertebrate auditory system. Hearing Research 12:31-55

Fitzgibbons PJ (1984) Temporal gap detection in noise as a function of frequency, bandwidth and level. J Acoust Soc Amer 74:67-72

Frisch K von (1938) Über die Bedeutung des Sacculus and der Lagena für den Gehörsinn der Fische. Z Vergl Physiol 25:703-747

Frisina RD, Smith RL, Chamberlain, SC (1984) Responses to amplitude modulation in the cochlear nucleus: A hierarchy of enhancement. J Acoust Soc Amer 75:S68

Furukawa T, Hayashida Y, Matsuura S (1978) Quantal analysis of the size of excitatory post-synaptic potentials at synapses between hair cells and afferent nerve fibers in goldfish. J Physiol 276:211-226

Furukawa T, Kuno M, Matsuura S (1982) Quantal analysis of decremental response at hair cell-afferent fiber synapses in the goldfish sacculus. J Physiol 322:181-195

Furukawa T, Matsuura S (1978) Adaptive rundown of excitatory post-synaptic potentials at synapses between hair cells and eighth nerve fibers in the goldfish. J Physiol 276:193-209

Green DM (1973) Minimum integration time, in Basic Mechanisms in Hearing (A Møller, Ed.). Academic Press, London, p 773

Hall L, Patricoski M, Fay RR (1981) Neurophysiological mechanisms of intensity discrimination in the goldfish, in Hearing and Sound Communication in Fishes (W Tavolga, A Popper, R Fay, Eds.). Springer-Verlag, New York, p 179

Hawkins AD (1981) The hearing abilities of fish, in Hearing and Sound Communication in Fishes (W Tavolga, A Popper and R Fay, Eds.). Springer-Verlag, New York, p 109

Hawkins A, Chapman C (1975) Masked auditory thresholds in the cod (Gadus morhua). J Comp Physiol 103:209-226

Henderson D (1969) Temporal summation of acoustic signals by the chinchilla. J Acoust Soc Amer 46:474-475

Jerger JF (1955) The influence of stimulus duration on the pure tone threshold during recovery from auditory fatigue. J Acoust Soc Amer 27:121-124

Johnson RA (1980) Energy spectrum analysis in echolocation, in Animal Sonar Systems (RG Busnel, JF Fish, Eds.). NATO Advanced Study Institute Series. Plenum, New York, p 673

Kuno M (1983) Adaptive changes in firing rates in goldfish auditory nerve fibers as related to changes in mean amplitude of excitatory postsynaptic potentials. J Neurophysiol 50:573-581

Langner G (1984) This volume

Leshowitz B, Wightman FL (1971) On-frequency masking with continuous sinusoids. J Acoust Soc Amer 49:1180-1190

Manning FB (1924) Hearing in the goldfish in relation to the structure of its ear. J Exp Zool 41:284-292

Menna D 81984) This volume

57

Myrberg AA, Jr, Spires JY (1972) Sound discrimination by the bi-
   color damselfish (Eupomacentrus partitus) J Exp Biol 57:727-735
Offutt GC (1967) Integration of the energy in repeated tone pulses
   by man and the goldfish. J Acoust Soc Amer 41:13-19
Platt C (1977) Hair cell distribution and orientation in goldfish
   otolithic organs. J Comp Neurol 172:283-297
Plomp R, Bouman A (1959) Relation between hearing threshold and
   duration for tone pulses. J Acoust Soc Amer 31:749-758
Popper AN (1972) Auditory threshold of the goldfish as a function
   of signal duration. J Acoust Soc Am 52:596-602
Popper AN (1977) A scanning electron microscopic study of the sacculus and lagena in the ears of fifteen species of teleost
   fishes. J Morph 153:397-418
Popper AN, Clarke NL (1978) Non-simultaneous auditory masking in
   the goldfish (Carassius auratus). J Exp Biol 83:145-158
Schmiedt R, Zwislocki J (1980) Effects of hair cell lesions on
   responses of cochlear nerve fibers. II. Single and two tone intensity functions in relation to tuning curves. J Neurophysiology
   43:1390-1405
Schuijf A (1981) Models of acoustic localization, in Hearing and
   Sound Communication in Fishes (W Tavolga, A Popper, R Fay, Eds.).
   Springer-Verlag, New York, p 267
Schuijf A, Hawkins AD (1983) Acoustic distance discrimination by
   the cod. Nature 302:143-144
Sinnott J (1984) Mechanisms in auditory discrimination: Ontogeny
   and phylogeny. J Acoust Soc Amer 75:S21
Tavolga W (1974) Signal/noise ratio and the critical band in
   fishes. J Acoust Soc Amer 55:1323-1333
Yost WA, Hill R, Perez-Falcon T (1978) Pitch and pitch discrimina-
   tion of broad band signals with rippled power spectra. J Acoust
   Soc Am 63:1166-1173
Zwicker E (1975) Scaling, in Handbook of Perception vol 5/2 (W
   Keidel, W Neff, Eds.). Springer-Verlag, New York, p 441
Zwislocki J (1969) Temporal summation of loudness: An analysis. J
   Acoust Soc Amer 46:431-441
Zwislocki J (1973) On intensity characteristics of sensory recep-
   tors: A generalized function. Kybernetik 12:169-183

Time Resolution in the Auditory Systems of Anurans

Robert R. Capranica, Gary J. Rose[1] and Eliot A. Brenowitz[2]

Section of Neurobiology and Behavior, Cornell University,
Ithaca, New York 14853, USA

INTRODUCTION

In order for an animal to detect an acoustic signal, its ear must
be sensitive to the frequency components in that sound. Any sounds
that fall outside its audiogram will be inaudible. This is very
obvious and has been the basis for most neuroethological studies of
animal sound communication. In general, auditory nerve fibers have
"V"-shaped frequency tuning curves and complex sounds produce
various excitatory patterns of this peripheral array of frequency
filters; the remainder of the frequency recognition process occurs
centrally (Capranica and Moffat 1983). There is no doubt that
studies of frequency processing in the auditory system have proven
valuable. But studies of frequency processing by themselves are inadequate to fully understand the neural basis of species-specific
acoustic communication. The simultaneous dimension of temporal processing must be included for such an understanding (Capranica and
Rose 1983). In fact it may well be that the encoding of temporal
features will be more crucial than frequency features in reaching
that goal.

Anurans are excellent models for neurobehavioral studies of temporal signaling. In many species, males call with rapid rhythmicity
and answer each other with remarkable timed precision (e.g., Littlejohn and Martin 1969; Loftus-Hills 1974; Narins and Capranica
1978). But these interactions involve gross temporal characteristics and their encoding can be easily understood on the basis
of onset and duration of overall excitation of the auditory system.
A more intriguing question concerns the fine-temporal structure in
acoustic signals. It is well known that anurans in a variety of
genera produce calls with specific pulse repetition rates and
amplitude modulation features (e.g., Blair 1958; Gerhardt 1978a;
Loftus-Hills and Littlejohn 1971; Nevo and Capranica 1984;
Walkowiak and Brzoska 1982). Their ability to produce rapid temporal modulations resides in specializations in their laryngeal
apparatus (Martin 1972; Schneider 1977). An example of the mating
calls of two species of toads, Bufo fowleri and Bufo americanus, is
shown in Figure 1. The carrier frequency in each species' call

[1]Neurobiology Unit, Scripps Institute of Oceanography, UCSD, La
Jolla, California 92093, USA
[2]Department of Psychology, University of California, Los Angeles,
California 90024, USA

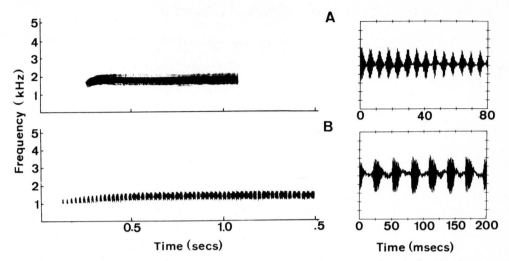

Fig. 1. Sound spectrograms (left) and oscillograms (right) of the mating calls of (A) Bufo w. fowleri and (B) Bufo americanus (Rose and Capranica 1984a)

corresponds to the natural vibration rates of their vocal cords. For these two species the carrier frequencies are rather similiar. The superimposed amplitude modulation is produced by separate (passive) vibration of the arytenoid cartilages which overlie the vocal cords, resulting in a trill. The temporal sequence of the amplitude-modulated pulses within the trill probably arises from the pulsation pattern of the thoracic musculature which is under neural control (active mechanism). The trill rate of Fowler's Toad is about 120 Hz whereas the American Toad has a much lower trill rate of approximately 40 Hz. Clearly the primary difference in the species specificity of the calls of these animals resides in their rates of amplitude modulation, and that brings us to the focus of our study. How does the peripheral and central auditory system process temporal rates such as these that are characteristic of many bioacoustic signals?

TEMPORAL CODE IN PERIPHERAL NERVOUS SYSTEM

Following to Click Trains

The anuran's inner ear contains two auditory organs which are tuned to separate species-specific frequency regions: the amphibian papilla is sensitive to low and mid frequencies whereas the basilar papilla is sensitive to a higher frequency range (Capranica 1976; Wilczynski and Capranica 1984). Their disparate frequency tuning reflects the anuran's peripheral specialization for selective detection of sounds of communicative significance. The eighth-nerve fibers that innervate these organs fire in synchronization to acoustic stimuli that have periodic waveforms. Such an example to trains

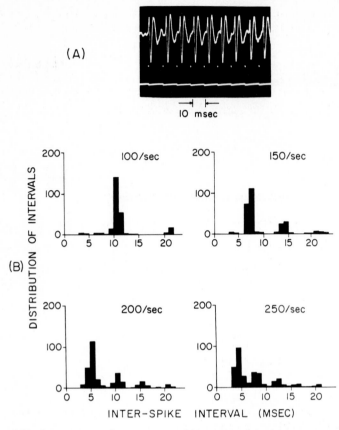

Fig. 2. (A) Response of a nerve fiber in Scaphiopus couchi to a train of clicks. The lower trace indicates the occurrence of each click (rate 95/sec, click duration 0.2 msec); the upper trace shows the unit's ability to fire a single spike to each click in the train. (B) Typical interspike interval histograms for the same fiber in response to clicks at various repetition rates. Distinct peaks in the histogram indicate preferred firing times and correspond to multiples of the interclick interval (Capranica and Moffat 1975).

of broad-band clicks is shown in Figure 2 for the Spadefoot Toad Scaphiopus couchi. For low repetition rates the spikes are time-locked in a one-to-one fashion to the individual clicks in the train. As the click repetition rate increases, the interval between successive spikes decreases and the fiber begins to show evidence of skipping some of the clicks. Nevertheless for click rates up to at least 250/sec, there is clear evidence that auditory nerve fibers can preserve the temporal periodicity in a stimulus by their precise interspike intervals. The trill rate in the mating call of Couch's Spadefoot Toad is approximately 150/sec and the repetitive pulse rate in its release call is about 50/sec. These repetition rates are well within the capabilities of its auditory nerve fibers to "follow" repetitive waveform events. Thus we find that, while each fiber may perform a filter analysis on the basis of its frequency tuning curve, it simultaneously encodes the temporal periodicities in that filtered output by its interspike intervals.

Phase-Locking to Tones

A similar example of the precision of this simple temporal code can be seen in the cycle histograms in Figure 3. These histograms show

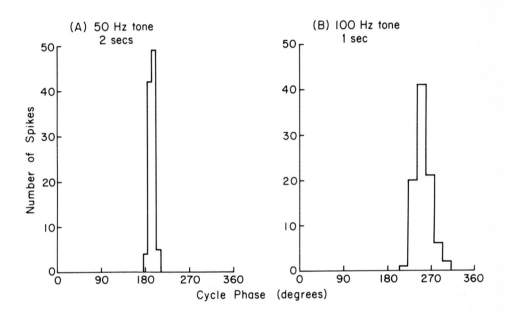

Fig. 3. Cycle histograms for an auditory nerve fiber in the American Toad Bufo americanus in response to (A) 50 Hz pure tone and (B) 100 Hz pure tone. The best excitatory frequency of this fiber was 115 Hz; tones presented at 20 dB above the fiber's threshold at each frequency. Bin width in both histograms is 0.5 msec.

the relative times of spikes during successive cycles of a low-frequency pure tone. The very narrow clustered distribution empha-sizes the ability of auditory nerve fibers in anurans to fire phase-locked spikes, thus conveying the periodicity in the stimulus by a simple interspike interval code. The accuracy of this code can be judged by the coefficient of synchronization within the cycle histogram (Goldberg and Brown 1969). A coefficient of unity indi-cates firing at the exact same moment in every cycle, whereas a value of zero indicates random firing times during each cycle. The coefficient of synchronization in Figure 3 to a 50 Hz tone is 0.96 and to a 100 Hz tone it is 0.88. Such high synchronization values again illustrate that auditory nerve fibers in anurans can code waveform periodicities with high precision. Narins and Hillery (1983), in their electrophysiological studies in the auditory nerve of the neotropical treefrog Eleutherodactylus coqui, have found that the degree of synchronization falls off monotonically with tonal frequency. Nevertheless significant phase-locked synchronization to tones extends to 900 Hz.

Synchronization to Amplitude-Modulated Noise

These results with pulse trains and tones indicate that an anuran's peripheral auditory system can encode temporal periodicities up to rather high rates. But periodic click trains and pure tones contain discrete spectral cues so that it is not so clear to what extent the frequency composition within these signals may influence temporal coding measurements. By using sinusoidally amplitude-modulated white noise as a stimulus set, the depth and rate of modulation can be varied without altering the spectral composition or the overall RMS sound pressure level (i.e., the spectrum remains flat). Thus amplitude-modulated noise is a preferred stimulus for studies of temporal processing. Surprisingly the response level of eighth nerve fibers in anurans is largely independent of the rate of amplitude modulation, as can be seen for two representative fibers in Figure 4.

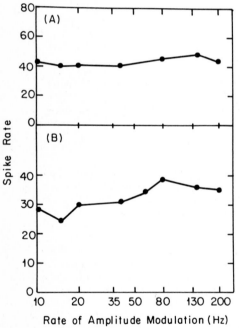

Fig. 4. Spike rate (spikes/sec) vs rate of amplitude modulation of white noise for two eighth-nerve fibers in the leopard frog Rana pipiens. The neuron in (A) had a best excitatory frequency to tones of 140 Hz and was stimulated with 100% AM white noise at 67 dB SPL; the neuron in (B) had a best excitatory frequency of 820 Hz and was stimulated with 100% AM white noise at 49 dB SPL (Rose and Capranica 1984b).

Each fiber's firing level remains fairly constant for AM rates over the range of at least 10 - 200 Hz which encompasses the usual rates in most bioacoustic signals. An increase in intensity simply leads to a rather uniform increase in firing level over this entire AM range. Based on systematic recordings from 125 auditory nerve fibers in leopard frogs (Rana pipiens), Rose and Capranica (1984b) conclude that there is no selective temporal tuning in the periphery to any preferred AM rate. This stands in marked contrast to the selective frequency tuning curves of auditory nerve fibers. Rather, as in the case of responses to click trains and pure tones, eighth nerve fibers faithfully preserve all modulation rates over the range of 10 - 200 Hz by their synchrony of firing during successive cycles of the modulating envelope. A representative example can be seen in Figure 5. The discharges are phase-locked to the modulating waveform.

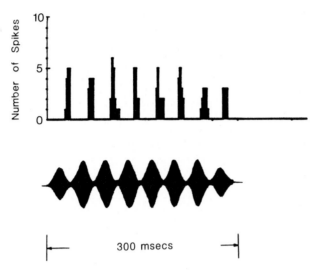

Fig. 5. Peri-stimulus-time (PST) histogram demonstrating a clear synchronization of an auditory nerve fiber's spikes in the leopard frog to a particular phase of the modulating waveform. The stimulus was white noise amplitude modulated (100%) at a rate of 38 Hz (Rose and Capranica 1984b).

The precision of a fiber's phase-locked activity to sinusoidally-modulated white noise can be described by a coefficient of synchronization between 0 and 1 which reflects the distribution of firing times during successive cycles of the AM envelope, a measure very similiar to the synchronization to pure tones. As the depth of modulation is increased, there is a fairly uniform increase in synchronization over the entire range of AM rates of 10 - 200 Hz (Rose and Capranica 1984b). An increase in the overall intensity of the stimulus likewise leads to an increase in the coefficient of synchronization, but again it is a uniform increase over the entire range of AM rates. So, as in the case of firing level rates vs. AM rate, the coefficients of synchronization of eighth-nerve fibers do not exhibit a tuned selectivity to preferred rates of amplitude modulation. This is a very important point to appreciate. While the peripheral auditory system of anurans shows selective tuning to species-specific frequency regions, it does not exhibit selective tuning to temporal features. Figure 6 summarizes the coefficients of synchronization for 43 auditory nerve fibers in the leopard frog in response to 100 Hz amplitude-modulated (100%) white noise when compared to their best excitatory frequencies in response to pure tones. In general, fibers that are most sensitive to tones below

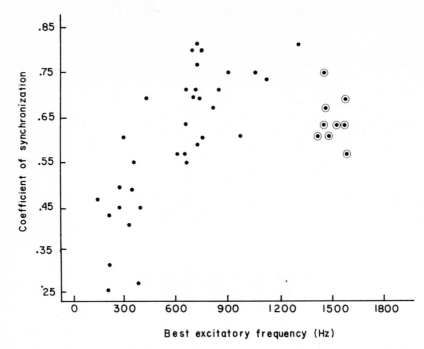

Fig. 6. *Relationship between best excitatory frequency to tones and coefficient of synchronization to white noise 100% amplitude modulated at 100 Hz for 43 nerve fibers in the leopard frog. Circled data points represent nerve fibers presumed to innervate the basilar papilla; uncircled points represent fibers that likely derive from the amphibian papilla (Rose and Capranica 1984b).*

400 Hz have lower synchronization values then fibers tuned to higher frequency tones. But for fibers with very similar best excitatory frequencies, there is a rather wide range in their coefficient of synchronization values. This suggests that even in the periphery there might begin to be a separation between frequency and temporal encoding in the auditory nervous system.

TEMPORAL CODE IN CENTRAL NERVOUS SYSTEM

Selectivity for Amplitude Modulation Rates

Let's now turn to the central nervous system and ask whether the simple periodicity code in the periphery is maintained throughout the ascending auditory system. The answer is that a transformation in this code occurs in several different ways. Rose and Capranica (1984b) have defined five distinct response classes in the leopard frog's midbrain (torus semicircularis) based on spike rate functions to amplitude-modulated white noise (Figure 7). Approximately 35% of the cells are "AM non-selective", namely their spike rates

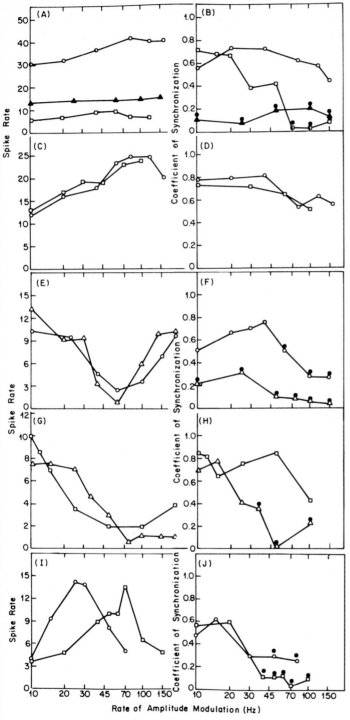

Fig. 7. Spike rate (spikes/-sec) vs rate of AM functions (A,C,E,G,I) and coefficient of synchronization vs rate of AM functions (B,D,F,H,J) for 11 representative neurons in the torus semicircularis of the leopard frog. Five AM response types are represented: AM non-selective (A,B); AM high-pass (C,D); AM band-suppression (E,F); AM low-pass (G,H); and AM tuned (I,J). All stimuli were 100% amplitude-modulated white noise at 10-20 dB above each neuron's threshold. The coefficient of synchronization reflects the degree of concentration of each unit's spikes to a particular phase of the modulation cycle histogram (bin width is 0.1 msec). All values were computed on the basis of at least 10 stimulus presentations. Non-significant synchronization values are indicated by the filled circles (Rose and Capranica 1984b).

are relatively independent of the rate of modulation. In this respect they resemble the uniform firing patterns of eighth nerve fibers, although their coefficients of synchronization can be quite different (Figure 7A,B). A second "AM high-pass" group (approximately 9%) shows a pronounced increase in firing level as the rate of amplitude modulation increases, but this change in activity level is not reflected in their synchronization values (Figure 7C,D). "AM band-suppression" units (about 9%) exhibit a reduction in firing to intermediate AM rates compared to lower or higher rates (Figure 7E,F). A fourth group (approximately 17%) responds to low AM rates but not to higher modulation rates and are termed "AM low-pass" neurons (Figure 7G,H). And finally the "AM tuned" group (about 30%) of the cells in the torus responds maximally to a rather narrow range of modulation rates (Figure 7I,J). For all five classes of cells, their responsiveness to amplitude-modulated signals cannot be predicted from their coefficient of synchronization functions. The ubiquitous firing synchrony that characterizes the peripheral auditory system is no longer present in the central auditory system. In other words the simple periodicity code in the auditory nerve has been transformed into a distribution of temporal selectivities for which interspike intervals no longer need represent the temporal fine-structure in an acoustic waveform. This is an important transformation in appreciating how complex sounds are represented centrally.

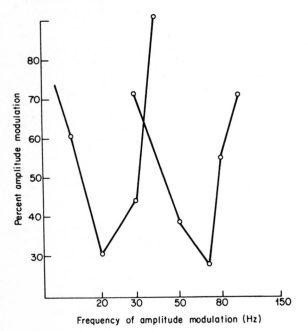

Fig. 8. Temporal tuning curves of two auditory units in the torus semicircularis of the leopard frog. One unit is tuned to 20 Hz AM, and the other to 70 Hz AM. In these measurements the stimulus intensity was held constant while the depth of modulation was varied to maintain a threshold criterion. At 100% AM the envelope of the noise varies sinusoidally from maximal amplitude to zero amplitude; for lower percentages of AM these excursions decrease so that zero percent AM corresponds to unmodulated white noise (Rose and Capranica 1983).

The AM-tuned neurons are especially interesting. Their selectivity to amplitude modulation can be characterized by an isoresponse temporal tuning curve (Rose and Capranica 1983). These tuning curves reflect the depth of modulation at different rates of AM required to maintain a threshold number of spikes, as shown in Figure 8. Their form is remarkably similar to a frequency tuning curve with a "V"-shape and a clear best rate of AM, and their sharpness of tuning can be characterized by a $Q_{20\%}$ value (best modulation rate divided by the bandwidth of its AM tuning rate at a modulation depth 20% greater than threshold; Rose and Capranica 1984b). Representative $Q_{20\%}$ values range from about 0.8 to 4.0. Given that these neurons can have both a frequency tuning curve and a temporal tuning curve, we might ask whether they are related. The surprising answer is that there is no obvious correlation between a cell's frequency selectivity and its temporal selectivity, as can be seen in Figure 9 on the basis of 25 units in the torus of the leopard frog. Thus units with similar best excitatory frequencies to tones can have quite different best rates of amplitude modulation.

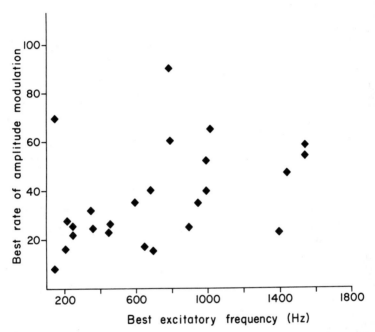

Fig. 9. Best rate of amplitude modulation vs best tonal excitatory frequency for 25 units in the torus of the leopard frog (Rose and Capranica 1984b).

Species-Specific Temporal Filters

What is the significance of cells that are selective for particular rates of amplitude modulation? We believe they play an important role in processing communication signals. As an example, let's return to the mating calls of Fowler's Toad and the American Toad in Figure 1. As we have already pointed out, the trill rate is about 120 Hz in Bufo fowleri, which is much higher than the rate of 40 Hz in Bufo americanus. What about the temporal tuning curves in

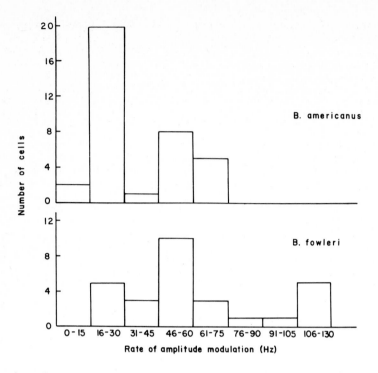

Fig. 10. Distribution of best rates of amplitude modulation for AM-tuned units in the torus semicircularis of (upper) American Toad and (lower) Fowler's Toad (Rose and Capranica 1984a).

these two species? Figure 10 shows the distribution of their best rates of amplitude modulation. There are several cells tuned to high rates between 90 - 130 Hz AM in Bufo fowleri. But these high rates are not represented in Bufo americanus; instead we find the largest number of cells are tuned to a much lower range of about 16 - 30 Hz. In fact in Figure 10 the median best rate for Bufo americanus is 26.0 Hz compared to 56.5 Hz for Bufo fowleri which is a very significant difference (p<0.0002, Mann-Whitney U-Test). In addition both species produce release calls which contain pronounced AM features (Rose and Capranica 1984a). In general, the rates in the release calls of Bufo fowleri are higher than those of Bufo americanus, but in both species there is a wide span of AM rates in these calls. And there likely are other sounds of biological significance for each species that contain temporal modulations. So a distribution of cells with different best rates would be expected.

Temperature Effects

Since auditory nerve fibers in anurans do not exhibit preferences for particular rates of amplitude modulation, then this tuning must be an intrinsic operation within the central nervous system. That is, temporal selectivities at the level of single cells in the brain must arise from specific neural connections, latencies and synaptic signs of activity. Given that spike conduction times and synaptic delays vary with temperature, we might expect that temporal tuning would shift with body temperature. Anurans, being poikilothermic, provide ideal subjects to test this prediction. We selected the gray treefrog Hyla versicolor for these studies because it is well known that the trill rate in the male's call increases linearly with temperature (Blair 1958; Gerhardt 1978b).

For example, at 16°C the trill rate is about 15 pulses/sec whereas at 24°C it increases to 24 pulses/sec. Furthermore a female prefers the calls of males corresponding to her own body temperature. By means of two-choice phonotaxis tests, Gerhardt (1978b) verified that females at 16°C prefer calls with pulse rates of 15/sec compared to 24/sec, whereas at 24°C this preference is reversed. Thus there is temperature coupling between the vocal system and the temporal pattern recognition system. If AM-tuned cells in the torus are involved in temporal pattern recognition, as we suggest, then they should show a shift in best rates with temperature that parallels these behavioral results. Such a shift would provide the neural basis of temperature coupling of acoustic signal reception to signal production.

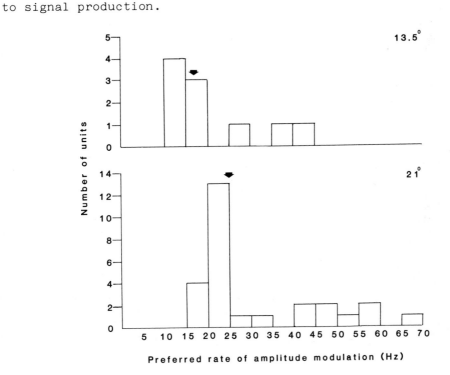

Fig. 11. Distribution of preferred rates of amplitude modulation of auditory units in the midbrain of the gray treefrog Hyla versicolor at a body temperature of 13.5^{o}C (top) and 21°C (bottom). Arrows indicate the median modulation rate for each distribution (Brenowitz, Rose and Capranica 1984).

In recordings from 118 single cells in the midbrain of 10 gray treefrogs collected in Ithaca, New York approximately one third were AM-tuned cells, each exhibiting a clear selectivity for a best rate of modulation. In conducting these experiments the animals were placed on a Peltier plate so that their body temperature could be precisely controlled (and continuously monitored by means of an electronic thermometer with a calibrated probe placed deep into the cloaca). Figure 11 shows the distribution of preferred rates to amplitude-modulated white noise for animals at $13.5^{\circ}C$ compared to animals at $21.0^{\circ}C$. There is an obvious difference in the distributions at these two temperatures ($Z = -5.91$, $p < 0.05$, Mann-Whitney U-Test). The median best rate of AM for the units recorded at $13.5^{\circ}C$ is 17 Hz whereas at $21.0^{\circ}C$ the median best rate is 25 Hz. Thus as we anticipated at higher temperatures cells in the torus of the gray treefrog respond to higher rates of modulation. To verify this conclusion unequivocally, in three cases we were able to record continuously from the same single cell as the animal's body temperature was slowly increased from $13.5^{\circ}C$ to $21.0^{\circ}C$. These experiments are a tour de force and are not meant for the impatient! But alas, perseverance prevailed and Figure 12 shows an example of

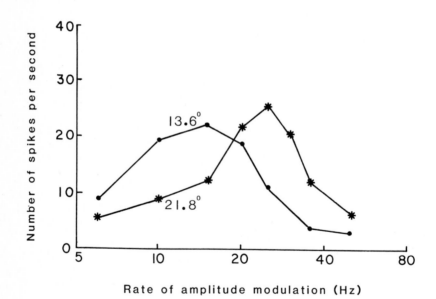

Fig. 12. Shift in AM tuning of a representative cell in the torus semicircularis of the gray treefrog as a function of temperature: $13.6^{\circ}C$ (o -- o) and $21.8^{\circ}C$ (x -- x) (Brenowitz, Rose and Capranica 1984).

the fruits of these labors. At $13.6^{\circ}C$ the cell exhibits maximal firing to AM rates around 15 Hz. However at $21.8^{\circ}C$ its preferred rate has shifted to 25 Hz. The direction and magnitude of this shift closely follows the change in trill rate in the male's call, as well as the concomitant change in a female's behavioral selectivity to this same temporal feature over this temperature range. The correlation is quite remarkable. We found similar shifts in tuning in the other two cells that were tracked continuously during a systematic temperature change. These experiments were so convincing that they leave no question in our minds about the existence of temporal tuning at the level of single cells in the central nervous system and the importance that this tuning must play in processing acoustic signals of significance, including temperature coupling between the auditory and vocal systems of cold-blooded vertebrates.

CONCLUDING REMARKS

Cells in the central auditory system that are tuned to temporal features should be anticipated, since the time domain is an equivalent alternative to processing in the frequency domain (Capranica and Rose 1983). In addition to our studies, evidence is beginning to emerge in other laboratories that filters for temporal fine-structured details likely exist at the level of single cells (Bibikov and Gorodetskaya 1980; Rees and Møller 1983; Walkowiak 1980). Since there is a distribution of such temporal filters, as we have found, it may be difficult to identify them individually by gross recording techniques (Hillery 1984). Most of us are used to the notion of frequency filters and Fourier analysis provides the basis of our routine spectral description. But what about temporal filters? How should we proceed in identifying them in the course of our systematic electrophysiological studies of the auditory system, especially with regard to an animal's behavior. What should we look for? It's quite likely that those dimensions are selective and species-specific just like the frequency domain. If that indeed is the case as we suspect and have suggested, then many of us face an interesting challange. Our experimental design was based on a fairly clear motivation for studies of a relatively simple temporal feature in a model bioacoustic vertebrate. The design of similar experiments and their interpretation in more advanced species such as birds and mammals, which have a much greater acoustic diversity, poses a real challenge. Nevertheless we hope our studies in anurans provide some guidelines for those future efforts.

ACKNOWLEDGEMENTS

This work was conducted at Cornell University under support of the U. S. National Institutes of Health (NINCDS Grant NS-09244).

REFERENCES

Bibikov NG, Gorodetskaya ON (1980) Single unit responses in the
    auditory center of the frog mesencephalon to amplitude-modulation
    tones. Neirofiziologiya 12:264-271 (English translation in Neuro-
    physiol 12:185-195, 1980)
Blair WF (1958) Mating call in the speciation of anuran amphibians.
    Amer Nat 92:27-51
Brenowitz EA, Rose GJ, Capranica RR (1984) Neural correlates of
    temperature coupling in the vocal comunication system of the gray
    treefrog (Hyla versicolor). Submitted to Science
Capranica RR (1976) Morphology and physiology of the auditory sys-
    tem. In Frog Neurobiology, Llinás R and Precht W, eds. Springer-
    Verlag, Berlin, pp 551-575
Capranica RR, Moffat AJM (1983) Neurobehavioral correlates of sound
    communication in anurans. In Advances in Vertebrate Neuroethol-
    ogy, Ewert J-P, Capranica RR, and Ingle DJ, eds. Plenum, London,
    pp 701-730
Capranica RR, Rose GJ (1983) Frequency and temporal processing in
    the auditory system of anurans. In Neuroethology and Behavioral
    Physiology, Huber F and Markl H, eds. Springer-Verlag, Berlin, pp
    136-152
Gerhardt HC (1978a) Mating call recognition in the green treefrog
    (Hyla cinerea): the significance of some fine-temporal proper-
    ties. J Exp Biol 74:59-73
Gerhardt HC (1978b) Temperature coupling in the vocal communication
    system of the gray treefrog, Hyla versicolor. Science 199:992-994
Goldberg JM, Brown PB (1969) Response of binaural neurons of dog
    superior olivary complex to dichotic tonal stimuli: some phy-
    siological mechanisms of sound localization. J Neurophysiol
    32:613-636
Hillery CM (1984) Detection of amplitude-modulated tones by frogs:
    Implications for temporal processing mechanisms. Hearing Research
    (in press)
Littlejohn MJ, Martin AA (1969) Acoustic interaction between two
    species of leptodactylid frogs. Anim Behav 17:785-791
Loftus-Hills JJ (1974) Analysis of an acoustic pacemaker in Stre-
    cker's Chorus Frog (Anura: Hylidae). J Comp Physiol 90:75-87
Loftus-Hills JJ, Littlejohn MJ (1971) Pulse repetition rate as the
    basis for mating call discrimination by two sympatric species of
    Hyla. Copeia 1971:154-156
Martin WF (1972) Evolution of vocalization in the genus Bufo. In
    Evolution in the Genus Bufo, Blair WF, ed. Univ Texas Press,
    Austin, pp 279-309
Narins PM, Capranica RR (1978) Communicative significance of the
    two-note call of the treefrog Eleutherodactylus coqui. J Comp
    Physiol 127:1-9
Narins PM, Hillery CM (1983) Frequency coding in the inner ear of
    anurans and amphibians. In Hearing - Physiological Bases and
    Psychophysics, Klinke R and Hartmann R, eds. Springer-Verlag,
    Berlin, pp 70-76
Nevo E and Capranica RR (1984) Evolutionary origin of ethological
    isolation in cricket frogs, Acris. Evol Biol (in press)
Rees A, Møller AR (1983) Responses of neurons in the inferior col-
    liculus of the rat to AM and FM tones. Hearing Research 10:301-331

Rose GJ, Capranica RR (1983) Temporal selectivity in the central auditory system of the leopard frog. Science 219:1087-1089

Rose GJ, Capranica RR (1984a) Processing amplitude-modulated sounds by the auditory midbrain of two species of toads: matched temporal filters. J Comp Physiol 154:211-219

Rose GJ, Capranica RR (1984b) Sensitivity to amplitude modulated sounds in the anuran auditory nervous system. J Neurophysiol (in press)

Schneider H (1977) Acoustic behavior and physiology of vocalization in the European tree frog Hyla arborea (L.). In The Reproductive Biology of Amphibians, Taylor DH and Guttman SI, eds. Plenum, New York, pp 295-335

Walkowiak W (1980) The coding of auditory signals in the torus semicircularis of the fire-bellied toad and the grass frog: Responses to simple stimuli and to conspecific calls. J Comp Physiol 138:131-148

Walkowiak W, Brzoska J (1982) Significance of spectral and temporal call parameters in the auditory communication of male grass frogs. Behav Ecol Sociobiol 11:247-252

Wilczynski W, Capranica RR (1984) The auditory system of anuran amphibians. Progress Neurobiol 22:1-38

Aspects of the Neural Coding of Time in the Mammalian Peripheral Auditory System Relevant to Temporal Resolution

E.F. Evans

Department of Communication & Neuroscience,
University of Keele, Staffs, ST5 5BG, U.K.

INTRODUCTION

Many psychophysical investigations in man (reviewed elsewhere in this volume) and in animals (Giraudi et al 1980) indicate that the mammalian auditory system is capable of discriminating discontinuities in the amplitude of stimuli occuring in intervals as short as a few milliseconds. In "gap-detection" tasks (e.g. Plomp 1964; Fitzgibbons and Wightman 1982) gaps, of the order of 3-5 ms, can be discriminated between stimuli typically of a few hundred milliseconds' duration, in broad band or high frequency stimuli at high sensation levels. The minimum gaps detected become progressively longer at lower sensation levels, particularly below 25-30 dB SL, approaching 25 ms at 10 dB SL. The latter authors found an inverse relationship between threshold gap duration and the frequency of the (octave) band used as the stimulus.

It is generally agreed that the physiological substrate for auditory temporal resolution is predominantly, though not exclusively, within the auditory periphery. Generally implicated is a "decay in sensation", presumably correlating with some physiological property outliving the duration of the physical stimulus. Duifhuis (1973) has suggested that an early component of temporal masking (less than 10 ms) may be related to the decay of the response of the peripheral auditory filter preceding the cochlear nerve fibres. On this basis, temporal resolution should be related to the frequency of the stimulus and the band-width of the peripheral auditory filter concerned. In a constant "Q" system (constant relative band-width) the duration of the impulse response of the filter will decrease with increasing centre frequency of the filter. This is consistent with the direction of the results obtained by Fitzgibbons and Wightman (1982). In contrast, however, in sensory-neural hearing loss of cochlear origin, the band-width of the cochlear filters increases (see Evans 1978b for review), but the expected improvement in temporal resolution does not occur; in fact it deteriorates (e.g. Jesteadt et al 1976; Fitzgibbons and Wightman 1982).

In this brief review, evidence will be presented that the properties of the mammalian peripheral auditory system are optimal in two respects for the purposes of temporal resolution. Firstly, the nature of the peripheral cochlear filters are such that an optimal balance appears to have been struck between temporal and spectral resolution. Secondly, neural mechanisms of "off-suppression" at the level of the cochlear nerve and "off-inhibition" at the level of

the cochlear nucleus, serve to limit the discharge of neurones
following the cessation of stimuli, thereby enhancing the temporal
contrast. Lastly, observations will be made on the effects of
impairment of cochlear function on the coding of temporally discontinuous stimuli and possible effects on neural "off-suppression".

Detailed reviews of the extensive literature on the temporal coding
of simple and complex stimuli in the auditory periphery have been
made elsewhere (Evans 1975, 1978a, 1981).

PERIPHERAL AUDITORY FILTERS

All modern investigations of the tuning of mammalian cochlear nerve
fibres show them to be remarkably sharply tuned (see Evans 1975,
1978b for review). The equivalent rectangular (approximately half-
power) band-widths of the cochlear fibre filters for narrow and
broad band signals average about 100-200 Hz for fibres with characteristic frequencies (CF) up to about 1 kHz, and about 10% of the
CF from 2 kHz upwards. The cut-off slopes of these filter functions
are extremely steep, particularly for fibres with CFs above about 2
kHz where they exceed 50-200 dB/oct. High resolution measurements
of these filter functions show them to be "U" rather than "V"
shaped (see Fig. 2A). Using the reverse-correlation technique of de
Boer and Kuyper (1968), the impulse response of the (linear part of
the) cochlear filter can be determined (Figs. 1 and 2; see also
Evans 1977). Fourier transform of these impulse response functions
yield attenuation functions which match the neural frequency threshold curves (FTCs) remarkably well (Fig. 2A). The shape of the FTCs
and the form of the impulse responses strongly suggest that the
cochlear filtering process resembles more a multiple band-pass
filter than a simple resonant filter. A comparison of the impulse
responses and attenuation functions of such filters is shown in
Fig. 3. Here the filter characteristics of a simple resonant filter
(dashed line and upper impulse response) and a multiple band-pass
filter (continuous line and lower impulse response), having approximately equal durations of the impulse response are compared. The
multiple band-pass filter achieves substantially steeper cut-offs
than the resonant filter at the expense of a somewhat wider
half-power band-width. The multiple pole band-pass filter utilized
in Fig. 3 is in fact a one third-octave filter, and the form and
duration of its impulse response match reasonably well the impulse
responses of Fig. 1. To achieve an equivalent frequency resolution
in terms of the steeper cut-off skirts, a resonant filter would
have to have a substantially longer impulse response, and therefore
poorer temporal resolution.

The impulse response functions derived from the reverse correlation
analyses indicate that the filtering represented by the pure tone
FTC is reflected in the frequencies dominating the temporal patterns of discharge, on which the analysis depends. In other words,
the discharge pattern of the cochlear nerve fibre tends to follow
preferentially frequencies within the effective band-width of the
fibre (up to the frequency limit of phase-locking of about 4-5
kHz). Fig. 2 also shows that the form of the impulse response

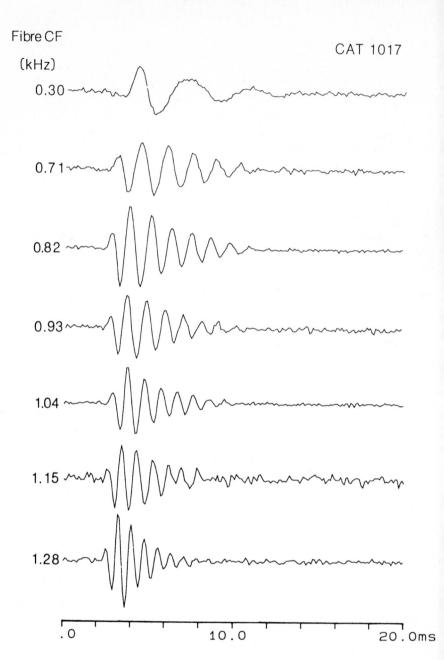

Fibre CF (kHz)

CAT 1017

0.30
0.71
0.82
0.93
1.04
1.15
1.28

.0   10.0   20.0ms

Fig. 1. Impulse responses of 7 cochlear nerve fibres filters in the same cat, derived by reverse correlation of their discharges in response to 4 kHz low-pass filtered gaussian noise at about 40 dB above threshold. Method of Evans (1977) used, except that a true reverse-correlation paradigm is employed here, i.e. all spikes used for analysis. Each impulse response results from averaging 20 ms of the noise stimulus immediately preceding each discharge spike. The delay before the onset of the impulse response is arbitrary but consistent for all fibres. Bin width 100 µsec.

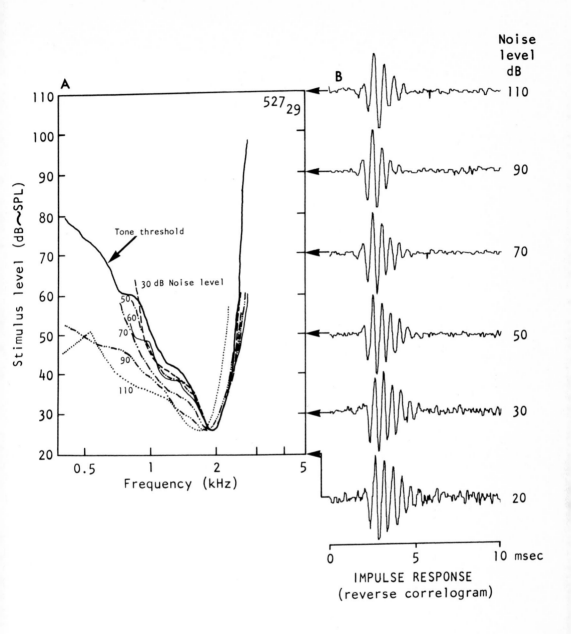

Fig. 2. Filtering characteristics (power spectrum and impulse response), derived by reverse correlation at different noise stimulus levels, for a cochlear nerve fibre with CF of 2 kHz. A: continuous line: frequency threshold curve obtained with pure tone stimulation. Interrupted lines: Fourier transforms of the impulse responses (B) obtained by reverse correlation of the spike discharges with the broad-band noise stimulus at the noise level indicated. (From Evans 1977).

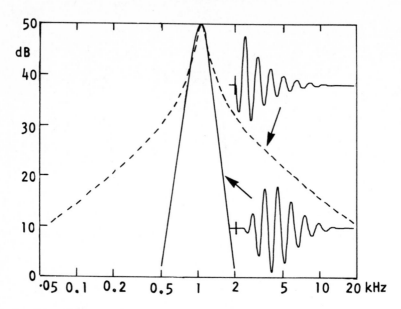

Fig. 3. Comparison of the time and frequency responses of resonant and multiple pole band-pass filters having similar durations of impulse response. (See text).

functions is maintained with little modification over a very wide dynamic range. This means that irrespective of the stimulus level, the temporal patterning of cochlear fibres is dominated by the frequencies in the stimulus weighted by the cochlear filter function indicated by the FTC, i.e. largely by the frequencies within the effective (half-power) bandwidth. Deviations from this situation occur at high sound levels, as shown in Fig. 2. For cochlear fibres with CFs above about 1 kHz, the band-width of the cochlear filter increases, and the low frequency cut-off slope decreases with stimulus level, particularly above 30-40 dB above threshold. In the cat, where these deviations are relatively minor (compared with the rat: Møller 1978, and guinea pig: Harrison and Evans 1982), the effective band-widths increase by a factor of between 1.1 and 2.5 at the highest noise levels used (100-110 dB SPL) and this is accompanied by a small but systematic shift downwards in CF with level (Fig. 2A). For fibres with CFs below about 1 kHz, the shift in the CF is towards higher frequencies (Evans 1977).

These properties are also reflected in the temporal pattern of activity of cochlear nerve fibres to stimuli having multiple frequency components (Evans 1980a, 1981), as shown in Fig. 4. Here the cochlear fibre is being stimulated with a 10 harmonic (equal amplitude) complex centred on its CF (1.4 kHz), the fundamental frequency being 200 Hz. The degree to which the temporal discharge patterns synchronize with each harmonic ("phase-locking") is plotted as the "vector strength" against the frequency of the harmonic in the lower part of the figure on a logarithmic scale so that comparison may be made with the pure tone FTC. At stimulus levels within 40 dB of threshold, the frequency weighting of the vector

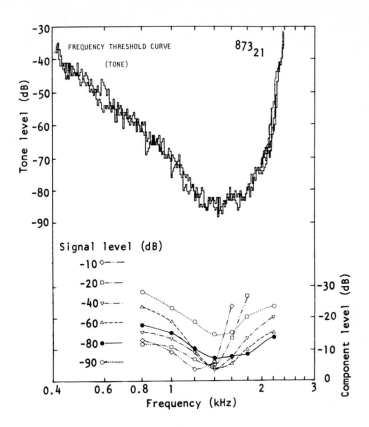

Fig. 4. Comparison of weighting functions for the phase-locking response of a cochlear fibre to individual components of a harmonic complex, compared with the fibre's frequency threshold curve. Upper curve: automatically derived frequency threshold curve for a cochlear fibre with characteristic frequency of 1.4 kHz. Lower curves: relative vector strengths (a measure of degree of phase-locking) of the response to the individual harmonics of a 10 harmonic tone complex presented at different stimulus levels. 0 dB = 106 dB SPL for the complex as a whole. Vector strengths are plotted on a log power scale for comparison with the frequency threshold curve. At stimulus levels near threshold (-90 dB e.g.) the weighting functions are similar to the FTC. At the highest signal levels (e.g. -10 dB: curve joining the diamond points), there is a shift of the centre frequency of the filter function downwards (to 1.2 kHz), a reduction in the slope of the low frequency cut-off and an increase in the slope of the high frequency cut-off. (From Evans 1981).

strengths reflects the FTC. At higher stimulus levels, similar changes in selectivity are reflected in the weighting of the relative vector strengths as described above in the case of reverse correlation (Evans 1981). For this fibre, having a CF above 1 kHz (1.4 kHz), the half-power band-width increases slightly, the low frequency cut-off slope decreases and the centre frequency of the filter shifts downwards (curve joining diamond shaped points). For fibres with CFs below about 1 kHz, the CF of the weighting function shifts towards higher frequencies.

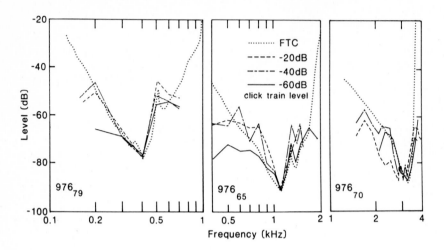

Fig. 5. Comparison of weighting functions derived from Fourier transforms of autocorrelation histograms in response to a click train (see text) for 3 fibres with characteristic frequencies of 0.37, 1.08 and 3.15 kHz respectively at 3 stimulus levels in each case, with the pure tone frequency threshold curve of each fibre. FTCs indicated by dotted lines; weighting functions: -60 dB level (about 10 dB above response threshold): continuous line; -40 dB level: dashed-dotted line; -20 dB level (about 50 dB above threshold): dashed line. (From Evans 1983).

Similar results have been obtained with impulsive stimuli (Evans 1983), as shown in Fig. 5. This shows the weighting, by the cochlear nerve filtering process, of a fibre's response to a click train having uneven intervals. The weighting functions are derived by FFT of the autocorrelation histograms of the discharge pattern evoked by the click train (see Evans 1983 for details). The figure shows the correspondence between the weighting functions obtained at three sound levels of the click train, compared with the pure tone FTC (dotted outline). The minor deviations between the weighting functions and the FTC are consistent with those encountered with broad band noise and the harmonic complexes as described above. At near threshold stimulus levels, the weighting function obtained with harmonic complexes and click trains is sometimes slightly wider than the FTC. At stimulus levels up to about 40-60 dB above threshold, however, the correspondence between FTC and weighting functions is close, particularly for fibres with CFs below about 1.5 kHz. At higher stimulus levels, the weighting function tends to shift towards lower frequencies for fibres with CFs above about 1 kHz (Fig. 5, unit 976.70) together with a progressive decrease in the slope of the low frequency cut-off.

These results indicate that, to a first approximation, the temporal response properties of cochlear nerve fibres can be considered to arise by analogy with a linear band-pass filter having the spectral and phase characteristics represented by the impulse responses of Figs. 1 and 2. This statement of course ignores the second-order effects described above due probably to synchrony-suppression (e.g. Rose et al 1974; Young and Sachs 1979). In synchrony suppression, typically the lower frequency stimulus components tend to dominate the temporal responses of cochlear nerve fibres, hence accounting for the shift in the filter centre frequency towards lower frequencies and the decrease in the low frequency cut-off slope noted above, for fibres with higher CFs.

As far as temporal resolution is concerned, however, these second order effects should not be very significant. To a first approximation, therefore, cochlear nerve fibres act for broad band and multi-component stimuli as if they were linear filters having the filter characteristics described by their pure tone FTC and their impulse responses as shown in Figs. 1 and 2. Given a fibre's FTC, therefore, its temporal dicharge properties in response to simple and complex sounds can therefore be reasonably well predicted and simulated by simple electronic analogues (Evans 1975, 1980b). The duration of the impulse response of the cochlear fibre filters, is, as Figs. 1 and 2 show in the cat, of the order of 11 ms for fibres with CFs in the 0.3 kHz region, decreasing to about 8 ms for CFs about 1 kHz (Fig. 1), and to about 4 ms for fibres with CF about 2 kHz (Fig. 2). These values are, in fact, comparable to those obtained in human psychophysical measurements of gap detection by Fitzgibbons and Wightman (1982). Our auditory temporal resolution could well be limited by the impulse response times of the cochlear filters, as suggested by Diufhuis (1973).

The above data were obtained from cochlear fibres in normal cochleas. It is well known that in a variety of pathological conditions of the cochlea their tuning characteristics deteriorate so that the band-widths of the cochlear fibre FTCs increase substantially (see Evans 1978b for review). The influence that this change in bandwidth has on the temporal patterning of cochlear nerve fibre discharges is illustrated in Fig. 6. The continuous line curve in the centre of the figure represents the FTC of a cochlear fibre from a cat cochlea in normal physiological condition, having a CF of 1.2 kHz. The inset histograms in the left hand corner of the figure, marked ISIH and AUTO, are respectively the interspike interval histogram and the autocorrelogram of that fibre in response to a click train having unevenly spaced alternate intervals of 4.7 and 5.3 ms (Evans 1983). This stimulus generates a complex of odd and even order harmonics of different amplitudes. Which of the harmonics "get through" the cochlear filter in order to interact and dominate the cochlear fibre temporal discharge patterns, determine the form of the interspike interval histogram and autocorrelogram. This can be illustrated, on the basis of the principles outlined above, by using a simple electronic analogue of the cochlear fibre (Evans 1980b) preceded by a linear filter having the same attenuation function as indicated by the FTC. The results of this simulation are shown in the autocorrelogram and interspike interval histogram in the top left hand corner of the figure. There is excellent correspondence between the output of the model and the

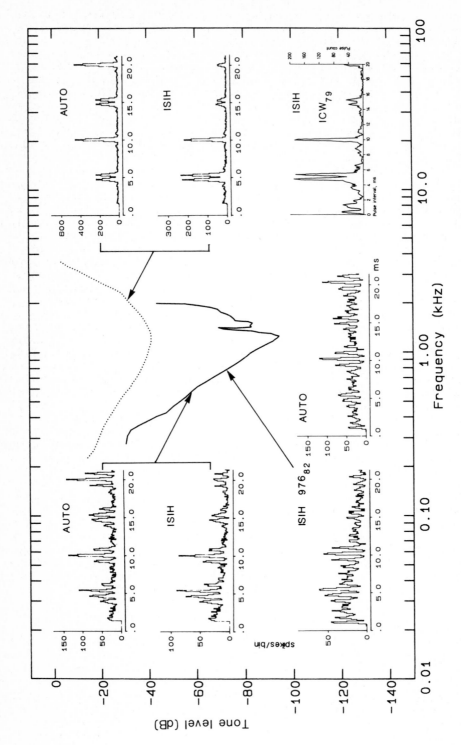

Fig. 6. Influence of band-width on temporal patterning of cochlear fibre discharges to multicomponent stimuli. (See text). (Data from Evans 1983).

real cochlear fibre data. This simulation was carried out original- ly, in order to test the author's contention (Evans 1983b) that data obtained by Whitfield (1979, 1980) using the same click train stimulus could not have been obtained with the sharp cochlear filtering associated with cochleas in normal physiological con- dition. The interspike interval histogram obtained by Whitfield is shown in the bottom right hand corner of the figure. This shows two strong peaks at 4.7 and 5.3 ms corresponding to the intervals in the stimulus. In order to model this interspike interval histogram satisfactorily, the electronic analogue had to be preceded by a filter function shown by the dotted FTC towards the top of the figure. This FTC was obtained from an experiment on a guinea pig of Harrison and Evans 1982, in which the cochlear condition was pathological as a result of chronic administration of kanamycin. Similar pathological effects can be obtained easily in the guinea pig as a result of deterioration in the physiological condition of the animal. Furthermore, in the guinea pig, the animal used in Whitfield's study, there is a stronger dependence of filter band- width on broad band stimulus level (Harrison and Evans 1982; Møller 1977). The FTC chosen from the experiment of Harrison and Evans was from a pathological cochlea and was derived using the reverse correlation technique with a high level noise stimulus probably comparable with the level of Whitfield's stimulation with the click train. Substitution of this FTC for the FTC from the normal cat cochlea produced from the model, autocorrelograms and interspike interval histograms (top right corner of the figure) virtually indistinguishable from those obtained by Whitfield (bottom right hand corner of the figure). This, indirectly, illustrates the importance of filter band-width in determining how many and which of the stimulus components will be sufficiently passed by the filter function to interact and therefore determine the temporal discharge patterns of the cochlear fibre. It also suggests that cochlear fibres, having wider bandwidths either resulting from cochlear pathology (as is likely in this case), or because of spe- cies differences (the chinchilla for example has wider band-widths than the guinea pig or cat - cf. Harris and Dallos 1979 with Evans 1972), exhibit greater temporal resolution, as would be expected. In Fig. 6, the stimulus periods are clearly resolved in the dis- charge pattern of the fibre with (estimated and simulated) broad tuning. That this is in the opposite direction to changes in psy- chophysical temporal resolution in patients with cochlear hearing impairment suggests that the deterioration in psychophysical tem- poral resolution may have little to do with cochlear filter charac- teristics per se. This discrepancy between physiological and psy- chophysical findings will be returned to in a later section.

"OFF-SUPPRESSION" AT COCHLEAR NERVE AND "OFF-INHIBITION" AT COCHLEAR NUCLEUS LEVELS

It has long been known that the cessation of stimulation of coch- lear nerve fibres is immediately followed by a transient reduction in the spontaneous activity and the responsiveness to further sti- mulation (see Evans 1975 for review, and the detailed studies by

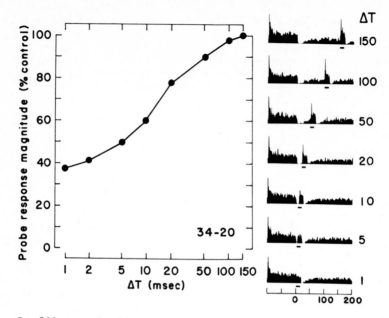

Fig. 7. Effects of off-suppression on the spontaneous discharge and evoked responses of a cochlear nerve fibre. Right hand part of figure shows poststimulus time histograms of response of cochlear fibre to a CF tone 30 dB above threshold, terminating at time 0 on the abscissa. The adapting response is followed by a period of complete suppression of the spontaneous activity and an exponential return to the previous spontaneous rate. Superimposed upon the latter is the response to a 15 ms probe tone (at the CF and 20 dB above threshold), commencing at the time indicated (ΔT ms) after the cessation of the adapting stimulus. With reduction in ΔT, a progressive reduction in the amplitude of the response to the probe tone is seen, particularly affecting the onset of the response. The magnitude of the probe response as a percentage of the control (unmasked) response is indicated in the left of the figure as a function of ΔT. (From Harris and Dallos 1979).

Young and Sachs 1973 for long-term stimulation and of Smith 1977, and of Harris and Dallos 1979 for analogous studies with short term stimulation). Fig. 7 shows both of these effects. Immediately after the cessation of a 100 ms tone at the CF of the fibre, at the point indicated by zero in the abscissa of the right hand part of the figure, a depression of the spontaneous activity occurs which, as can be seen at the top of the figure, lasts for about 10 ms before the spontaneous discharge rate returns exponentially to its spontaneous level. When a second (15 ms) tone (in this case again at the CF but 10 dB below the level of the first tone) is introduced, and the time interval between the cessation of the first tone and the onset of the second is progressively reduced (indicated by the ΔT values to the right of the figure) it will be seen that the response to the second tone diminishes as the interval is made shorter and the second tone lies in the period corresponding to reduction in the spontaneous discharge rate. The degree of this reduction in response is indicated in the left part of the figure as a percentage of the control value (i.e. in the absence of the first tone). The magnitude of the suppression of responsiveness and the time course of its recovery under various conditions have been carefully studied by Young and Sachs (1973); Smith (1977) and by Harris and Dallos (1979).

The magnitude of the suppression is a relatively simple function of the duration of the first (adapting tone) and the discharge rate evoked by it. Thus, as shown in Fig. 8, the magnitude of the

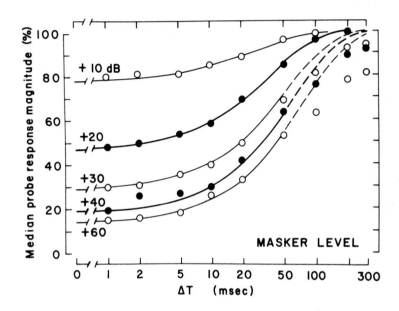

Fig. 8. Relationship between magnitude and time course of cochlear nerve off-suppression with level of the first (adapting/masker) stimulus. The response to the probe tone is indicated as a percentage of the response to the probe tone alone (unmasked response) as a function of the delay of the probe tone following the cessation of the first (adapting/masker) tone i.e. ΔT. With increasing level of the masking tone (parameter) the magnitude of the off-suppression and its time constant increases with stimulus level. All stimuli at CF. (From Harris and Dallos 1979).

suppression increases with the level of the first stimulus. This increase in the magnitude of the off-suppression follows simply the sigmoid relation between discharge rate and stimulus level (the rate level function) for the fibre. Likewise, increasing the duration of the first stimulus increases the magnitude of the "off-suppression" as shown in Fig. 9, the increase in magnitude of suppression being indicated by the progressively lower values where the curves meet the ordinate at zero ΔT.

The time constant of recovery of responsiveness from this off-

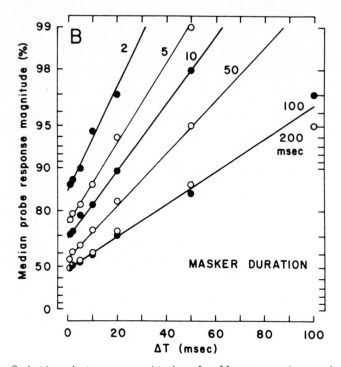

Fig. 9. Relation between magnitude of off-suppression and the time constant of its recovery with masker duration. As in the previous two figures, the magnitude of the probe response relative to its unmasked value is plotted against its time of occurrence after the cessation of the first (adapting/masking) tone. The parameter is the duration of the first (adapting/masking) tone. The ordinate is expressed logarithmically to illustrate the change in time constant (slope of lines). Increase in masker duration increases the magnitude of off-suppression (value of ordinate at ΔT = 0) and the time constant (slope) of the recovery from the off-suppression. The time constants for the curves from above down are 12, 14, 18, 23 and 37 ms respectively. (From Harris and Dallos 1979).

suppression has been studied by the above authors. As with the magnitude of off-suppression, its time course of recovery depends on the level and duration of the first (adapting) stimulus. The time course of recovery increases with both level and duration at least, in the findings of Harris and Dallos, with durations below 100 ms (Fig. 9). The same figure indicates that durations of the first (adapting) stimulus between 100 and 200 ms gave no significant difference in time constant. The time constants obtained increased from 12 ms for first stimulus durations of 2 ms, to 37 ms for stimulus durations between 100 and 200 ms.

Besides this dependence of the time constant of recovery on the duration and level of the first tone, there appears also to be a species difference (Harris and Dallos 1979). The time constants obtained in chinchilla (the figures quoted above) are about one half of the values obtained for the guinea pig, and probably the gerbil, used by Smith (1977).

In contrast to the effects of level of the first (adapting/masking) stimulus, the level of the second (probe) tone has little effect on the magnitude and time-course of the off-suppression (Smith 1977). The combination of the effects of the level of the first and second stimuli is such that when the level of both is changed equally, the response to the second (probe) stimulus remains approximately constant relative to the response to the first stimulus (Fig. 10).

These post-stimulatory suppression effects are found even more strongly in the cochlear nucleus, where neural inhibition serves to prolong the period of off-suppression substantially (Evans and Nelson 1973; Evans 1975), see Fig. 11. In this example, the effect of the off-inhibition is sufficiently strong and cumulative that the effect of changing the repetition rate of a 0.5 sec stimulus from 1 per 3 sec (A) to 1 per 0.8 sec (B) is to eliminate entirely the spontaneous activity of the unit and to attenuate substantially the onset of the response.

These suppression and inhibitory effects at cochlear nerve and nucleus levels serve to enhance the temporal resolution of peripheral elements in the auditory system to rapidly changing stimuli. Fig. 12 shows the responses of a cochlear nucleus unit to frequency and amplitude modulated tones presented at 20 dB above threshold. Period histograms C, G and I are modulation histograms of the averaged firing density in response to sinusoidal frequency modulation over a band of frequencies encroaching into the upper half of the units FTC. The waveform of the modulation of frequency is indicated under each modulation histogram. Lower frequencies therefore correspond as would be expected to higher probabilities of discharge. The rate of sinusoidal frequency modulation was 1 cycle per second in C, 10 in G and 40 c/s in I respectively. To the left of each histogram is a short horizontal bar indicating the spontaneous discharge (i.e. in the absence of any stimulation). It will be seen that at the higher rates of modulation (10 and 40 c/s) that the "tails" of the roughly sinusoidal distribution of firing probability are reduced below the level of the spontaneous activity. Likewise similar effects are obtained from sinusoidal amplitude modulation of a tone at the CF of the unit, with modulation rates of 1, 10 and 40 c/s: F, H, J respectively. This means that the contrast between the responses to the peaks and the valleys of the modulation become enhanced at the higher rates of modulation.

Quantitative studies of this effect have been made by Møller (e.g. 1972, 1976) in the cochlear nucleus and nerve, see Fig. 13. Compared with cochlear nucleus responses, the depth of modulation of the cochlear nerve fibre response largely reflects that of the stimulus as indicated by the equality (zero "dB gain") between the response and stimulus modulation depth. The relationship between response modulation and the modulation rate is essentially low-pass, with a high frequency cut-off primarily determined by the bandwidth of the cochlear fibre filter. This happens simply because at modulation rates numerically higher than half the bandwidth of the cochlear fibre response area, the sidebands in the stimulus fall outside the response area. Thus the stimulus components fail

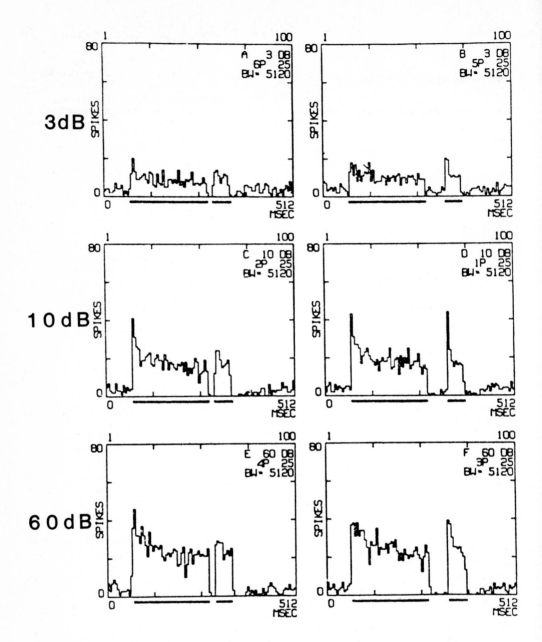

Fig. 10. Response of single cochlear nerve fibre to adapting and probe tones of equal level, 3, 10 and 60 dB above threshold respectively. Post-stimulus time histograms of response to adapting and probe tones at CF (1.8 kHz) indicated by long and short horizontal bars respectively under the histograms, separated by 16 ms silent gap in left-hand column, and 54 ms in the right-hand column. (From Smith 1977).

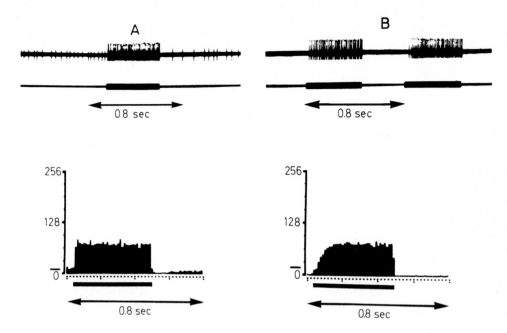

Fig. 11. Effects of "off-inhibition" in responses of a single unit in the dorsal cochlear nucleus of the anaesthetized cat. Upper half of the figure shows continuous film records of responses to a 0.5 sec tone at CF, 20 dB above threshold. In (A) the repetition of the tone is 1 per 3 sec; in (B) 1 per 0.8 sec. Below the continuous film records are the peri-stimulus histogram averages of the responses before, during and following the stimulus tone (horizontal bar) for the two conditions. Note in the case of the shorter repetition period, loss of normal spontaneous activity (indicated by short horizontal bar near the histogram ordinate) and the attenuation of the onset of the response to the tone. (From Evans 1975).

to interact within the response area, and therefore amplitude modulation of the stimulus component falling within the fibre response area does not occur. Since the response area bandwidth of cochlear nerve fibres is related to their characteristic frequency (CF), the low-pass cut-off frequency of the modulation transfer function will be a function of CF. For fibres with CFs up to about 1 kHz, the cut-off frequency would be expected to be of the order of 50-100 Hz; at a CF of 10 kHz, it will be about 500 Hz (as in Fig. 13 where the CFs of fibre and cell respectively were 7.8 kHz and 10.7 kHz). For many cells in the cochlear nucleus however, the modulation of discharge is enhanced by the off-inhibitory mechanisms described above so that the depth of the response modulation exceeds that of the stimulus over a wide range of modulation frequencies. This is seen in the cochlear nucleus cell of Fig. 13 (heavy line) as a "gain" exceeding 0 dB. The optimal modulation frequency is typically between 50-200 Hz. The optimal modulation frequency will obviously be a function of the time-constants of the off-inhibitory process.

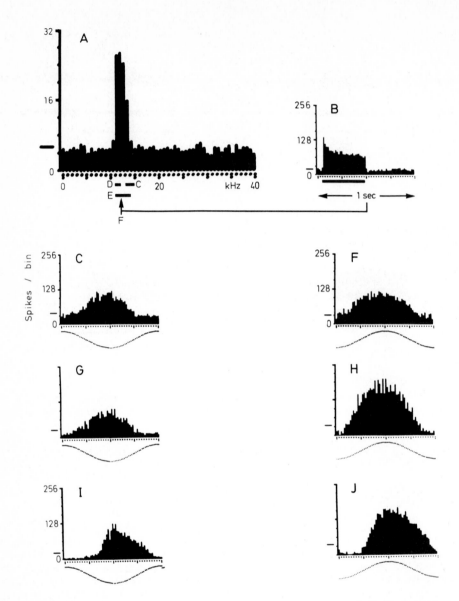

Fig. 12. Enhancement of temporal resolution by effects of "off-inhibition" in cochlear nucleus. A: frequency response histogram of response to 100 ms tones as a frequency of frequency. B: peri-stimulus time histogram of response to tone burst (horizontal bar) at CF. C, G, I: modulation histograms of averaged firing density in response to sinusoidal frequency modulation over band of frequencies indicated by bar C under frequency response histogram in A. Waveform of modulation indicated under modulation histograms. Modulation rate: 1, 10, 40 c/s in C, G, I respectively. F, H, J: amplitude modulation (depth 80% at CF; CF indicated by F in A) at modulation rates of 1, 10 and 40 c/s respectively. At the higher rates of modulation (G, I and H, J) note depression of "tails" of distributions below spontaneous level (indicated by short bars against ordinates). All analyses at 20 dB above threshold and data collected for about 30 sec in each case. (From Evans 1975).

psychoacoustic gap thresholds and the level of the stimuli, which relationship may not be inconsistent with the physiological data (e.g. Fig. 10). In the psychophysical data of Plomp (1964), the gap detection threshold duration varied from about 25 ms at 10 dB SL to about 4 ms at 75 dB SL. In the physiological data, increase in the level of both pre- and post-gap stimuli has little effect on the responses, but the contrast of the inter-response gap is heightened by the increase in off-suppression brought about by the increase in level of the pre-gap stimulus (Fig. 10). This heightened contrast should mean that the minimum gap for detection would reduce with level. Obviously, this conjecture must be checked with physiological studies directly addressed to the psychoacoustic gap-masking paradigm.

As far as comparisons between the physiological "off-suppression" and psychophysical forward masking are concerned, Harris and Dallos (1979) have discussed the similarities and differences in detail. Briefly, there appears to be a discrepancy in terms of the relationships between the level of the masker and the time course of the forward masking, and in the dynamic range of masker level effects. On the other hand, the influence of masker duration is similar in both cases, namely that increasing the masker duration beyond 100 ms has little effect on the magnitude of off-suppression and psychophysical forward masking whereas reducing the masker durations below about 100 ms causes a progressive attenuation of both effects.

TEMPORAL RESOLUTION IN HEARING IMPAIRMENT

A striking recent finding in psychophysical investigations of cochlear hearing impairment is a deterioration in temporal resolution measured in a variety of ways (Jesteadt et al 1976; Tyler et al 1982; Fitzgibbons and Wightman 1982). Gap detection thresholds, for example, are increased by a factor of one and a half to two (Fitzgibbons and Wightman 1982). This finding contrasts with surprising little evidence of gross changes in temporal coding at the cochlear nerve level in animals with pathological cochleas compared with normal function (Harrison and Evans 1979b; excepting the special case of salicylate poisoning, associated with abnormalities in the temporal distribution of intervals in spontaneous activity - Evans et al 1981). The gross form of the time course of adaptation and recovery (with the qualification below) in pathological conditions appeared to be relatively normal. Likewise, the relation between degree of discharge synchrony and stimulus frequency was not significantly different in pathological compared with normal cochleas, although the range of synchrony at any frequency was greater in the pathological case. Since the results of this study was published, a report by Woolf et al (1981) suggests that in pathological cochleas, the upper limit of phase-locking may be slightly but significantly reduced in pathological cochleas.

A closer examination of the data of Harrison and Evans (1979b), however, (Evans 1980c) suggests that in the pathological cochleas the time course of the recovery from the suppression may be

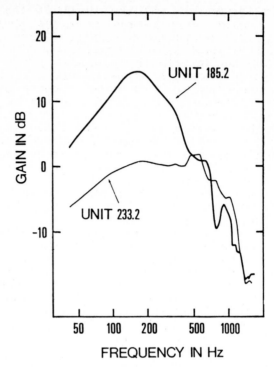

Fig. 13. Amplitude modulation transfer functions for cochlear nucleus cell (upper heavy line) and cochlear nerve fibre (thin line). Against the frequency of modulation is plotted the relative "gain" of the response, i.e. the ratio between the modulation of the histogram of the discharges and the modulation of the sound, expressed in dB. Data computed from cross-covariance functions derived from stimulation of the units with tones amplitude modulated by pseudorandom noise. (From Møller 1976).

RELATION BETWEEN PHYSIOLOGICAL DATA AND PSYCHOPHYSICAL MEASUREMENTS OF TEMPORAL RESOLUTION

While the mechanisms of off-suppression and off-inhibition must be relevant to establishing the limits on auditory temporal resolution, it seems that a simple correlation between these phenomena and temporal masking effects cannot be made in the present state of knowledge.

The mechanisms of off-suppression and off-inhibition are likely to be the result of neural phenomena and therefore would not be expected to be related to the CF of the unit. Thus no obvious relationship between the time course of the off-suppression and the CF of the unit has been found by Harris and Dallos (1979), in contrast to the relationship found in the psychophysical data by Fitzgibbons and Wightman (1982). However, there is a relationship between the extended compared with normal cochleas. This finding needs to be confirmed systematically, but it is in the same direction as the following reports. In the study by Young and Sachs (1973) of the time course of off-suppression following long (circa 60 sec) stimulation, it was noted that fibres with low spontaneous rates (e.g. below 30 spike/sec) had longer time constants of recovery from off-suppression. In the analogous study by Smith (1977) but with short duration (circa 200 ms) tones, it was noted in one animal, the cochlear sensitivity of which had decreased by about 6 dB in the course of the experiment, that many fibres were encountered whose time constants of recovery from off-suppression were "among the slowest observed" and the units had negligible spontaneous activity. It is not uncommon to observe a predominance of fibres having low or absent spontaneous activity in cochleas with both acute (e.g. Evans 1972, 1978b) and chronic pathology (e.g. Harrison and Evans 1979a).

It is a little difficult to predict the significance for perception of such effects on off-suppression in cochlear pathology. However, if the time course of the suppression is extended so that the onset of the neural response to the stimulus following the "gap" is attenuated, (as in the extreme case of Fig. 11), such an attenuation could be deleterious for discrimination of the gap. Obviously more work is required at the physiological and theoretical levels. It is, however, in line with the findings of Nelson and Turner (1980) indicating a prolonged recovery from forward masking in patients with cochlear hearing impairment.

Alternatively, the disruption in peripheral frequency representation occurring as a result of deterioration in cochlear frequency selectivity in cochlear impairment, may serve to affect deleteriously the processing of temporal cues by the central regions of the auditory system.

ACKNOWLEDGEMENT

I am grateful to the Medical Research Council for support for the cited studies.

REFERENCES

Boer E de, Kuyper GF (1968) Triggered correlation. IEEE Trans Biomed Eng 15:169-179

Diufhuis H (1973) Consequences of peripheral frequency selectivity for nonsimultaneous masking. J Acoust Soc Am 53:1471-1488

Evans EF (1972) The frequency response and other properties of single fibers in the guinea pig cochlear nerve. J Physiol London 226:263-287

Evans EF (1975) The cochlear nerve and cochlear nucleus. In Handbook of sensory physiology Vol V: pt 2. Eds. Keidel WD, Neff WD, Springer-Verlag, Heidelberg, 1-108

Evans EF (1977) Frequency selectivity at high signal levels of single units in cochlear nerve and cochlear nucleus. In Psychophysics and Physiology of Hearing, Ed. Evans EF, Wilson JP, New York: Academic Press, pp 185-192

Evans EF (1978a) Place and time coding of frequency in the peripheral auditory system: some physiological pros and cons. Audiol 17:369-420

Evans EF (1978b) Peripheral auditory processing in normal and abnormal ears: physiological considerations for attempts to compensate for auditory deficits by acoustic and electrical prostheses. Scand Audiol Suppl 6:9-44

Evans EF (1980a) "Phase-locking" of cochlear fibres and the problem of dynamic range. In International symposium on psychophysical, physiological and behavioural studies in hearing, Eds. vd Brink G, Bilsen F, Delft University Press, pp 300-309

Evans EF (1980b) An electronic analogue of single unit recording from the cochlear nerve for teaching and research. J Physiol 298:6-7P

Evans EF (1980c) Comment after Nelson & Turner "Decay of masking and frequency resolution in sensorineural hearing-impaired listeners". In International symposium on psychophysical, physiological and behavioural studies in hearing, Eds. vd Brink G, Bilsen F, Delft University Press, p 182

Evans EF (1981) The dynamic range problem: place and time coding at the level of cochlear nerve and nucleus. In Neuronal mechanisms of hearing, Eds. Syka J, Aitken L, Plenum Press, New York, pp 69-85

Evans EF (1983) Pitch and cochlear fibre temporal discharge patterns. In Hearing - physiological bases and psychophysics, Eds. Klinke R, Hartmann R, Springer-Verlag, Berlin, Heidelberg, New York, pp 140-145

Evans EF, Nelson PG (1973) The responses of single neurones in the cochlear nucleus of the cat as a function of their location and the anaesthetic state. Exp Brain Res 17:402-427

Evans EF, Wilson JP, Borerwe TAB (1981) Animal models of tinnitus. In: Tinnitus: CIBA Symposium 85:108-129. Pitman Medical, London

Fitzgibbons PJ, Wightman FL (1982) Gap detection in normal and hearing-impaired listeners. J Acoust Soc Am 72:761-765

Giraudi D, Salvi R, Henderson D, Hamernik R (1980) Gap detection by the chinchilla. J Acoust Soc Am 68:802-806

Harris DM, Dallos P (1979) Forward masking of auditory nerve fibre responses. J Neurophysiol 42:1083-1107

Harrison RV, Evans EF (1979a) Cochlear fibre responses in guinea pigs with well-defined cochlear lesions. In Models of the auditory system and related signal processing techniques, Eds. Hoke M, de Boer E, Scand Audiol Suppl 9:83-92

Harrison RV, Evans EF (1979b) Some aspects of temporal coding by single cochlear fibres from regions of cochlear hair cell degeneration in the guinea pig. Arch Otorhinolaryngol 224:71-78

Harrison RV, Evans EF (1982) Reverse correlation study of cochlear filtering in normal and pathological guinea pig ears. Hearing Research 6:303-314

Jestaedt W, Bilger RC, Green DM, Patterson JH (1976) Temporal acuity in listeners with sensorineural hearing loss. J Speech Hear Res 19:357-360

Møller AR (1972) Coding of amplitude and frequency modulated sounds in the cochlear nucleus of the rat. Acta Physiol Scand 86:223-238

Møller AR (1976) Dynamic properties of primary auditory fibers compared with cells in the cochlear nucleus. Acta Physiol Scand 98:157-167

Møller AR (1977) Frequency selectivity of single auditory nerve fibres in response to broadband noise stimuli. J Acoust Soc Am 62:135-142

Nelson DA, Turner CW (1980) Decay of masking and frequency resolution in sensorineural hearing-impaired listeners. In Psychophysical, physiological and behavioural studies in hearing, Eds. vd Brink G, Bilsen FA, Delft University Press, pp 175-182

Plomp R (1964) Rate of decay of auditory sensation. J Acoust Soc Am 36:277-282

Rose JE, Kitzes LM, Gibson MM, Hind JE (1974) Observations on phase-sensitive neurons of antero-ventral cochlear nucleus of cat: non-linearity of cochlear output. J Neurophysiol 37:218-253

Smith RL (1977) Short-term adaptation in single auditory nerve fibres: some poststimulatory effects. J Neurophysiol 40:1098-1112

Tyler RS, Summerfield Q, Wood EJ, Fernandes MA (1982) Psychoacoustic and phonetic temporal processing in normal and hearing-impaired listeners. J Acoust Soc Am 72:740-752

Whitfield IC (1979) Periodicity, pulse interval and pitch. Audiol 18:507-512

Whitfield IC (1980) Theory and experiment in so-called pulse-interval pitch. Audiol 20:86-88

Woolf NK, Ryan AF, Bone RC (1981) Neural phase-locking properties in the absence of cochlear outer hair cells. Hear Res 4:335-346

Young ED, Sachs MB (1973) Recovery from sound exposure in auditory nerve fibres. J Acoust Soc Am 54:1535-1543

Young ED, Sachs MB (1979) Representation of steady-state vowels in the temporal aspects of the discharge patterns of populations of auditory-nerve fibers. J Acoust Soc Am 66:1381-1403

Theoretical Limits of Time Resolution in Narrow Band Neurons

Dieter Menne

Lehrstuhl Zoophysiologie, Institut für Biologie III,
Auf der Morgenstelle 28, D-7400 Tübingen, F.R. Germany

ACCURACY AND RESOLUTION

When preparing the manuscript for this meeting on "Time resolution
in auditory systems" I looked through several textbooks on signal
processing and found at least three different meanings of the words
"time resolution". In the otherwise excellent book of Bendat and
Piersol (1980) it is used instead of "sampling period" in digital
signal processing; this is very unusual and should be avoided. The
second usage of the word "resolution" is to designate the precision
when measuring a time event. The apparent time delay between two
peaks of a signal corrupted by noise will not be the same when
measured repeatedly, but will show some statistical variance. At
low noise levels the variance will be low and therefore the measurement will be more accurate than at high noise levels. For describing this effect of statistical time jitter introduced by
noise, I will use the word "accuracy" instead of "resolution", as
it is usually done in literature on radar (Woodward 1955).

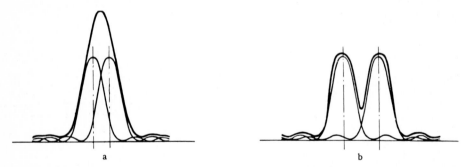

Fig. 1. Limits of resolution in the microscope. The separation of the
maxima must be larger than the half-width of the peak. (From Gerthsen
and Kneser 1969).

Following Woodward, I will adopt the third meaning of the word
"resolution", which best can be illustrated by the concept of resolution in microscopy as introduced by Rayleigh (Fig. 1). Viewed
through the microscope, each structure will produce a diffraction
pattern which is seen by the observer. If two points smaller than
the wavelength of the light are close to each other, their diffraction patterns overlap so that only one central peak is visible. It
now seems impossible to determine whether there are one or two
peaks: the two points can no longer be resolved. In an analogous
way, this definition could be used for time resolution. In contrast

to the definition of accuracy, it does not contain any reference to noise in the optical image or acoustic signal. Thus even without noise there seems to be a clear lower limit to the resolution of two objects. This apparent limit, however, can be overcome. In most cases, an algorithm or filter may be found that separates two unresolved peaks. If this is possible, the peaks are called resolvable. In the microscope example the algorithm could make use of the fact that the resulting peak of two objects is wider than the peak of one object.

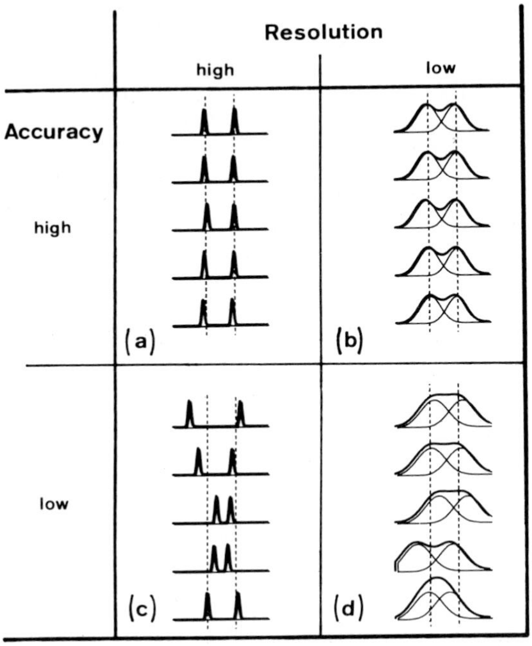

Fig. 2. Five samples of the output of a time measuring system with: (a) high resolution (sharp peaks) and high accuracy (little jitter) (b) low resolution (broad overlapping peaks) and high accuracy (c) high resolution and low accuracy (d) low resolution and low accuracy.

The concepts of accuracy and resolution are summarized in Fig. 2. Accuracy and resolution are independent concepts. A high resolution does not imply a high accuracy; a transformation which resolves two unresolved peaks will usually not improve, but possibly deterioate accuracy. Theoretical limits for the accuracy of time estimation as a function of the signal-to-noise ratio can be given quite exactly for technical systems (Woodward 1955, Burdic 1968) and have been applied to time estimation in the bat (Menne and Hackbarth, in press). In this paper I will discuss some problems in defining limits of resolution and their impact on biological research.

SMEARING AND RESHARPENING OF TIME PULSES

To demonstrate some aspects of resolution and resolvability, throughout this paper I will use a model which reflects some aspects of the vertebrate hearing system. This model is not intended to be a quantative description of the hearing system - the main conclusions will be qualitative and not dependent on the details of the model.

Fig. 3. Simplified model of one channel of an auditory system.

The first stage in the model (Fig. 3) is a narrow-band filter with a center frequency of 80 kHz and a very high Q-3 dB value of 600 (Q-10 dB = 200) (Fig. 4). The transfer function is somewhat similar

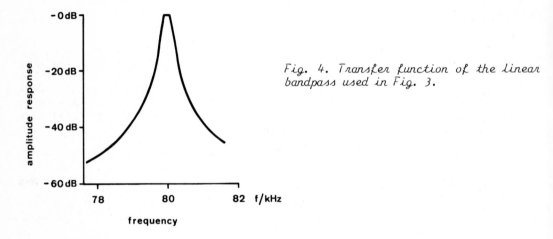

Fig. 4. Transfer function of the linear bandpass used in Fig. 3.

to the tuning curve of some neurons found in the peripheral audi- tory system of the bat <u>Rhinolophus ferrumequinum</u> (Suga et al. 1976). This somewhat exotic narrow-band neuron has been selected because it can be used to demonstrate some aspects of time resolu- tion. A system consisting of a bank of such filters with equispaced best frequencies would have a good frequency resolution: Two tones with only little frequency separation may excite different filters with at least one weakly excited filter between them.

The filtered signal is passed through an envelope detector. At the output of the envelope detector, the carrier frequency has been lost; so subsequent processing has to rely only on the envelope of the signal. This corresponds roughly to the situation found in the mammalian hearing system at frequencies above 5 kHz. The input frequency of the model is always assumed to be the center frequency of the model, 80 kHz. This is not an essential limitation but allows one to use only the envelope of the input signal in the figures.

If we apply a very short tone pulse (Fig. 5a) without any noise (Fig. 5b) to the input, the envelope at the output is considerably smeared out in time. This effect is a consequence of the narrow frequency response of the bandpass filter. In linear filters, the product of the bandwidth and the response time to an impulse is

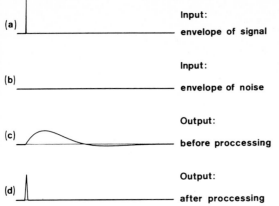

Fig. 5. (a) Envelope of a very short tone pulse of 80 kHz at the input of the model. (b) No noise is added. (c) Envelope of the pulse after passing through the bandpass. (d) Envelope sharpened with the filter of Fig. 8.

about 1. Although the envelope-detector is not a linear device, this relation is approximately valid for this model too; the 3 dB-bandwidth is roughly 100 Hz, the time constant 10 ms. The broadening in time is sometimes attributed to the "uncertainty principle of communication theory". This word has been adopted from quantum mechanics, where it really means that some parameters can only be determined with a certain probability. In communication theory the word is misleading, because there is no uncertainty associated with a broad time peak. If we had a device which exactly measured the peak position of the output pulse in Fig. 5, we would always get the same time value when repeating the measurement. Without noise, the error in time determination is zero even for a broad pulse, hence there is no limit for accuracy. Uncertainty will occur only when noise is added. The resulting error in time estimation is a monotonic function of the signal-to-noise ratio (SNR). At high SNR this timing error can be much less than the width of the output peak.

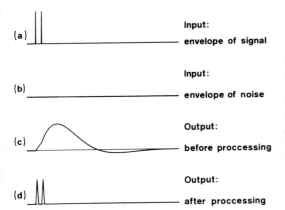

Fig. 6. (a) Envelope of two very short tone pulses of 80 kHz at the input of the model. (b) No noise is added. (c) The two peaks can no more be resolved in the envelope after passing through the bandpass. (d) The two peaks are resolved after sharpening with the filter of Fig. 8.

However, the smearing in time is a problem for time resolution. If two short tone bursts are applied to the model (Fig. 6), their output envelopes overlap and only one peak is visible - the two peaks are not resolved. In order to resolve the two peaks one has to find a transformation which sharpens the output envelope so that it looks similar to the input envelope. This is theoretically possible. Since the envelope of the output signal has a much lower frequency bandwidth than the envelope of the short pulse at the input, it is necessary to amplify the high frequency components of the envelope and to attenuate the low frequency components to obtain the sharpening. This type of sharpening is therefore called the spectral "whitening" approach; even though this term refers to the frequency domain, the processing can be carried out also using only the time signal. A technical realisation known as FIR (Finite Impulse Response) or nonrecursive filter is shown in Fig. 7 (Rabiner and Gold 1975).

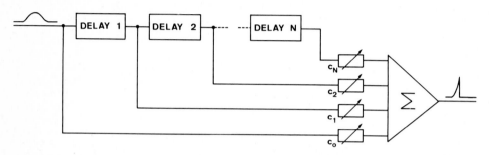

Fig. 7. Technical realisation of a finite impulse response (FIR, nonrecursive) filter.

A chain of delay elements stores a copy of the envelope or at least a section of it. The outputs of all delay elements are summed up with individual, but constant weights. By choosing the weighting coefficients suitably, one can obtain a wide range of filter characteristics, among these the desired whitening filter. Many delaying elements and corresponding coefficients are required in order to perform the sharpening operation with arbitrary precision. I have stripped down the design to only two delaying elements and three coefficients and will show that even with such a simple network the resolution can be remarkably improved.

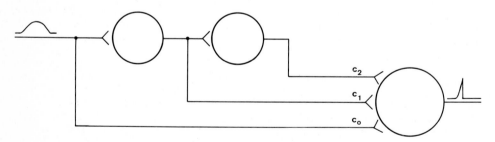

Fig. 8. Simplified finite impulse respone (FIR) filter used to sharpen the envelope using synaptic transitions as delay elements. A delay of 250 μs for each neuron is selected.

A possible way of realizing such a network with biological elements is shown in Fig. 8. A synaptical transition is used as a delaying element in this example, but since the delays can be quite short, the propagation delay along a neuron can be used as an alternate. For my illustrations, I have used a fixed delay of 250 μs per element. It is assumed that the output neuron carries out a weighted summation like the output element in Fig. 7 (this will certainly not be quite true in a real neuron). For the sharpening operation the center coefficient c_1 should have a negative sign (realizable by an inhibiting synapse). Using a gradient search technique I have found coefficients c_o, c_1, c_2 for the filter in Fig. 8, which maximally sharpen the smeared output pulse (Fig. 5c, d). With these coefficients two pulses at the input can be clearly resolved after processing (Fig. 6d). The reconstruction of the input envelope can even be carried further – the envelope of a signal like that in Figure 9a is restored quite accurately with this set of coefficients.

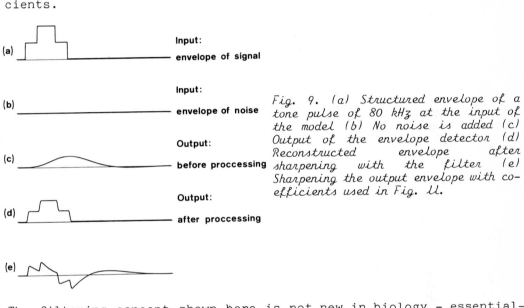

(a) — Input: envelope of signal

(b) — Input: envelope of noise

(c) — Output: before proccessing

(d) — Output: after proccessing

(e)

Fig. 9. (a) Structured envelope of a tone pulse of 80 kHz at the input of the model (b) No noise is added (c) Output of the envelope detector (d) Reconstructed envelope after sharpening with the filter (e) Sharpening the output envelope with coefficients used in Fig. 11.

The filtering concept shown here is not new in biology – essentially, it is equivalent to applying the concept of lateral inhibition to the time domain. So the following discussion can also be regarded as a discussion of the merits and drawbacks of lateral inhibition.

There are some serious drawbacks which limit the application of this sharpening method. One minor point comes from the fact that in my illustrations I have used a somewhat idealistic concept of envelope (calculated using the analytic signal) which still conserves quite a lot of fine grain information. With realistic envelopes as in neuronal systems the effect of sharpening will be much less pronounced, but still significant. More important is the fact that in the derivation of the sharpening filter no attempt has been made to take into account the influence of noise – the only objective was to compress the envelope in time. When broadband noise is added to the signal (Fig. 10), the output after processing is seriously contaminated by noise.

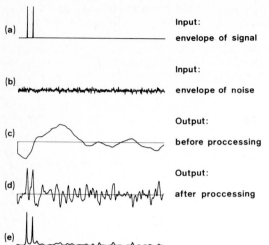

(a) Input:
envelope of signal

(b) Input:
envelope of noise

(c) Output:
before proccessing

(d) Output:
after proccessing

(e)

Fig. 10. (a) Envelope of two very short tone pulses of 80 kHz at the input of the model (b) Envelope of broadband noise added (c) Output of the envelope detector (d) Envelope after sharpening with the co-efficients used in fig. 5 (e) Output after processing of (d) with a high resolution algorithm.

The identification of the two peaks is still possible, but it is neccesary to know that only two peaks of about equal amplitude are expected and that the many other peaks therefore are "false alarms". This inclusion of a-priori knowledge which reduces ambiguity in resolution problems will be discussed in the next section.

Is it be possible to design a filter that sharpens the input and at the same time suppresses unwanted noise? This would be a filter which optimizes the signal-to-noise ratio at the output. With this constraint one gets the autocorrelation or matched filter, which can also be realized by a structure like in Fig. 7. For the example discussed here, a matched filter would not lead to an improved time resolution, since the improvement with such a filter is proportional to the time-bandwidth product of the envelope, which is about 1 for short signals after passing through a narrow band filter.

IMPROVING RESOLUTION WITH A-PRIORI INFORMATION

It has been shown that envelopes of time events can be restored quite accurately after the signal has passed through a narrow band system. The only thing the system has to know (or to learn) is the transfer characteristics of its input filtering system. But can it really be the only objective of a biological system to reproduce excellently the form of the input signal? Would it not rather try to interprete the input in a way which reflects its own previous experience? The towerlike envelope of Fig. 9 can be approximately reproduced by the processing, but this implies that the input envelope is interpreted by the system as one single structure - a possible alternative would be that it is regarded as the superposition of two rectangles of equal length (like in fig. 11a) which overlap in time. (There will be interferences if the carrier frequencies are not in phase, but we will assume that phase locking

can somehow be achieved). By choosing this alternative, we interprete the outer world; we attribute a higher probability to the
possibility "two overlapping rectangles" than to the alternative
"one tower". In this case one would not try to reconstitute the
input envelope, but rather process the envelope in such a way that
the two pulses are maximally separated in time.

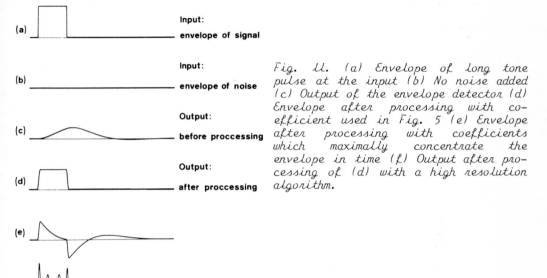

(a)   **Input:**
    envelope of signal

(b)   **Input:**
    envelope of noise

(c)   **Output:**
    before proccessing

(d)   **Output:**
    after proccessing

(e)

(f)

Fig. 11. (a) Envelope of long tone pulse at the input (b) No noise added (c) Output of the envelope detector (d) Envelope after processing with co-efficient used in Fig. 5 (e) Envelope after processing with coefficients which maximally concentrate the envelope in time (f) Output after pro-cessing of (d) with a high resolution algorithm.

The filter coefficients which perform this task can be found with
the same method as that used in Fig. 5. We take the output of the
envelope-detector to a broad pulse (Fig. 11c) as an input to the
filter structure of Fig. 8 and search for the coefficients which
maximally concentrate the output in time. The resulting coefficients will not reproduce the input pulse like in Fig. 11d, but
will perform a highpass filtering (Fig. 11e). If we use the new set
of coefficients to process the output of the towerlike signal of
Figure 9c, we obtain the interpretation we want: Two separate peaks
with positions corresponding to the beginning of the overlapping
sound pulses.

This is a somewhat oversimplified example which was constructed to
demonstrate that the limits of resolution cannot be defined as a
property of the system, but must be seen in conjunction with the
type of signals which it expects. I will now present a less trivial
method, which provides a more powerful tool for improving resolution in technical systems, but which is also more difficult to
understand and to realize. Most technical high-resolution methods
have been developed in order to obtain a better resolution in the
frequency domain - an excellent review is given by Kay and Marple
(1981). I have misused one of these methods (Marple 1980) to demonstrate high time-resolution by interchanging the time and frequency
domain in the original method.

The basic concept common to all high-resolution methods is to make a general assumption about the form of the signal and to fit a model to it. In our model it is assumed that the signal consists of a few peaks whose analytic form is given. It is essential that the fitting procedure should not explain all details of the signal, but allow for some residual error. The result of such a fitting procedure applied to the noisy double-pulse in Figure 10d is shown in Fig. 10e. The two "real" peaks are sharpened and the details of the noise are smoothed - a result which could not have been achieved with conventional filtering. However, if we use this type of processing for the broad pulse in Fig. 11d, the interpretation gets quite confusing. This is because the model assumes that the input structure consists of peaks, and it will therefore force this hypothesis on non-peaky structures. As long as the model corresponds to the input to the system, it really gives an improved resolution - if this assumption is wrong, the response may become worse than without processing.

RESOLUTION AND PATTERN INTERPRETATION

I do not think that Marple's method, nor any other numerical high-resolution method, can serve as a direct example of how animals can improve their time resolution, However, the electro-physiological experiments of Langner (1983) in the Guinea fowl demonstrate that similar methods are realized. The main concept of high-resolution methods can possibly be of some use for the inter-pretation of physiological data. It states that time resolution can be improved if we have some a-priori hypotheses about the types of signals which the system has to handle. The more the system "knows" about the possible alternatives, the higher is the improvement in resolution achieved. An objective system, which does not include information about the "acoustic biotope", cannot show optimal re-solving power. Since this a-priori knowledge cannot easily be for-mulated for a biological system, it is difficult to indicate a lower limit of resolution. On the other hand, it implies that the question of resolvability cannot be solved by a system theory ap-proach alone, but that it is a biological question which needs a detailed analysis of the role of the animal in its acoustic biotope.

It is a common idea in neuroethology that animals may make use of information about the signals which occur in their biotope. Inclu-sion of a-priori information in the sense of statistical in-formation theory corresponds directly to the specialization for processing "behaviorally relevant signals" in biology (Capranica and Rose 1983). By selecting the method to obtain a good time re-solution the animal performs a first step in pattern interpretation - thus the study of time resolution in more complex neurons can possibly serve as a model for simple pattern recognition.

A successful concept in neuroethology has been to look for neuronal structures which selectively respond to input signals important to the animal, e.g. conspecific frog calls. It has also been argued, that a neuronal system should selectively suppress interfering signals, e.g. that of calling frogs of another species. But what should such a system do with signals which have no relevance for it, because their a-priori probability is zero, e.g. the roaring of a lion for a North-American frog? Should it suppress this signal? Or can it afford to let it pass, since it normally does not lead to confusions? A technician asked to design such a system would write down all possible combinations of inputs and for each of them the answer "yes" (= respond) or "no" (= do not respond) at the output. With this truth table he has fully specified his system and can start to design the hardware. Such a system can be quite difficult to realize: for every input the corresponding output is fixed, there are no degrees of freedom left. But if the technician is sure that certain input combinations can never occur, then for this input he can write an "x" at the output, meaning "don't care". This term is somewhat misleading. It does not mean "do not respond at this input condition", but that the output condition is left open for these input. The system can now be built in such a way that the "don't care" outputs may be "yes" or "no" according to the convenience of the designer. Such a system with many "don't-care" cases often needs considerably less elements than a fully specified system. Applied to our resolution example it would mean that we can use a system like that in Fig. 10e to improve resolution, if we know that inputs like that in Fig. 11a (which produce false peaks at the output: Fig. 11f) can never occur. If an electrophysiologist uses long tone pulses to stimulate a neuron designed to treat such long pulses as "don't-care" cases, he will get a highly confusing result.

System which make use of "dont-care" cases can thus be build very economically; in real animals, the corresponding input condition will certainly be less trivial than in our example and more difficult to find if we use deterministic signals. It may be easier if noise is used as input signal. There are good arguments that noise is the ideal stimulus for system identification: it is a totally unbiased signal: every frequency and every sequence of frequencies occurs with equal probability. Using noise as a stimulus has been quite successful for analyzing the peripheral auditory system, e.g. when using the reverse correlation technique (de Boer and de Jongh 1978, Evans 1977). I think, however, that although second and higher order analysis of discharge statistics using noise at the input is theoretically appealing and gives nice pictures for complex neurons, the insight into the underlying functions that one obtains is remarkably small. The reason may be that highly specialized neurons have many "don't-care" cases. Using noise at the input, the neuron's proper biological function is also included in the response, but we do not see it, because it is completely buried by all the nonsense responses. If this is true, then noise as a most unbiased test signal would be a very bad signal for testing a system designed to meet only a limited set of situations. It would then not only be permissible to use biased signals known to be important for the animal, but we would even be forced to do so.

In this paper, I have limited my scope to time resolution in narrow-band neurons and tried to generalize some aspects to feature resolution. The same way of reasoning could have been used, if the words "time" and "frequency" had been interchanged - the title of the paper then would be: "Frequency resolution in neurons with short integration time". I think that the extreme view of discussing time versus frequency processing, is not realized in neuronal systems, and that it should therefore only be adopted for demonstration purposes. Most information can be extracted if time-frequency cells are taken together, and we should be ready to accept that animals make use of all this information.

SUMMARIZING THESES FOR DISCUSSION

1. The type of processing needed to obtain a good resolution can be different from the processing for optimizing the signal-to-noise ratio.
2. In order to achieve a good time resolution, it is advantageous to have a-priori knowledge about the types of signals which the system has to process.
3. When only a limited set of possible alternatives exists, the resolution can be improved by some type of model fitting which is a simple type of pattern interpretation.
4. If an animal only has to face a limited set of meaningful acoustic signals, in its biotope then evolution could treat the other situations as "don't-care" cases, thus simplifying the design of the neuronal network.
5. If a noise stimulus is used as input to a neuronal system which makes heavy use of "don't-care" cases, then the meaningful response could be totally buried by the majority of non-sense responses.

ACKNOWLEDGEMENTS

I thank Dr J Ostwald and Dr H-U Schnitzler for critical discussions.

REFERENCES

Bendat JS, Piersol AG (1980) Engineering applications of correlation and spectral analysis. John Wiley, New York, p 68

de Boer E, de Jongh HR (1978) On cochlear encoding: Potentialities and limitations of the reverse-correlation technique. J Acoust Soc Am 63:115-135

Burdic WS (1968) Radar Signal analysis. Prentice Hall, Englewood Cliffs

Capranica RR, Rose GJ (1983) Frequency and temporal processing in the auditory system of anurans. In Neuroethology and Behavioral Physiology, Huber F and Markl H, eds. Springer-Verlag, Berlin pp 136-152

Evans EF (1977) Frequency selectivity at high signal levels of single units in cochlear nerve and nucleus. In: Evans EF, Wilson JP (eds.). Psychophysics and physiology of hearing. Academic Press, London, pp 185-192

Gerthsen C, Kneser HO (1969) Physik. Springer-Verlag, Berlin Heidelberg New York, p 346

Kay SM, Marple Jr SL (1981) Spectrum analysis - a modern perspective. Proc IEEE 69:1380-1419

Langner G (1983) Evidence for neuronal periodicity detection in the auditory system of the Guinea Fowl: Implications for pitch analysis in the time domain. Exp Brain Res 52:333-355

Marple Jr SL (1980) A new autoregressive spectrum analysis program. IEEE Trans Acoust Speech Signal Process ASSP 28:441-454

Menne D, Hackbarth H (in press) Accuracy of distance measurement in the bat Eptesicus fuscus. J Acoust Soc Am

Rabiner LR, Gold B (1975) Theory and application of digital signal processing. Prentice Hall, Englewood Cliffs

Suga N, Neuweiler G, Möller J (1976) Peripheral auditory tuning for fine frequency analysis by the CF-FM bat Rhinolophus ferrumequinum. J comp Physiol 106:111-125

Woodward PM (1955) Probability and information theory with applications to radar. McGraw Hill, New York

Time Coding and Periodicity Pitch

G. Langner

Zoologie, Schnittspahnstrasse, 61 Darmstadt, FRGermany

INTRODUCTION: THE PITCH PROBLEM

A psychophysical attribute of periodic acoustic signals is pitch. All signals with the same period have the same pitch, irrespectively of their actual waveforms which are determined by the amplitudes and phases of their frequency components. For example, a sum of successive harmonics has the same period and elicits the same pitch as the fundamental (f_o) even if the lowest harmonic is much greater than f_o. This is the case of the "missing fundamental" and the percept was called "residue" (Schouten 1940a).

The residual and other psychophysical effects strongly suggest (de Boer 1976) that periodicity coding - especially neuronal time coding of the signal envelope - is the basis of pitch perception. However, prominent modern theories assume that the basis of pitch perception is a pattern analysis of a central representation of the spectrum (Terhardt 1972, Goldstein 1973, Wightman 1973). An advantage of such theories is that they could be based on spike rate given by the cochlear frequency analysis and be adjusted to fit certain pitch effects. However, it turns out that peripheral frequency resolution is not sufficient for that purpose (Sachs and Young 1979, Srulovicz and Goldstein 1983). Moreover, neuronal mechanisms suitable for the postulated pattern analysis are as yet unknown. It seems that the pitch problem is still open: is periodicity coded in the spectral or in the time domain, and how is it decoded?

The purpose of this paper is to discuss aspects of the psychophysical and the neurophysiological approaches to solve the pitch problem, including the author's correlation theory (Langner 1981, 1983a).

THE PSYCHOPHYSICAL APPROACH

Ohm (1943) suggested that the auditory system performs a Fourier analysis and resolves the frequency components of harmonic signals. Helmholtz (1862) used Ohm's "acoustical law" and introduced the place principle. According to this principle every frequency stimulates a distinct place on the basilar membrane and consequently elicits a distinct pitch perception. The pitch of harmonic signals composed of many frequencies is defined after Helmholtz by the lowest harmonic, while additional harmonics affect only the timbre.

In contrast, psychoacoustical experiments with a siren led Seebeck (1841) to the conclusion that the essential element for the perception of a pitch was the period of the signal's waveform, rather than it's fundamental. This observation was explained by Ohm and Helmholtz through the assumption that harmonic distortions generate the missing fundamental.

It was not until 100 years later, that Schouten (1940a) was able to rule out this explanation for Seebeck's pitch effect. Schouten produced harmonic signals without the fundamental frequency and added frequencies slightly different from the missing fundamental. The absence of any beating effects seemed to confirm Seebeck's point: the perceived pitch was not due to distortions, but to the total waveform due to harmonics unresolved by the auditory system.

Further evidence was obtained by Schouten using the amplitude modulation (AM) technique (Schouten 1940b). This allows to shift the frequencies of the individual components by the same amount, thereby keeping constant the frequency difference and as a consequence also the period of the envelope. Similar to the absence of beat effects, the observed pitch shift clearly contradicts the distortion hypothesis, which predicted a constant pitch evoked by the constant difference frequency of the signal components. These results were later confirmed by parametric studies of de Boer (1956). Surprisingly, since pitch shifted in spite of a constant envelope, these experiments at the same time ruled out Seebeck's assumption that pitch is readily defined by the envelope period.

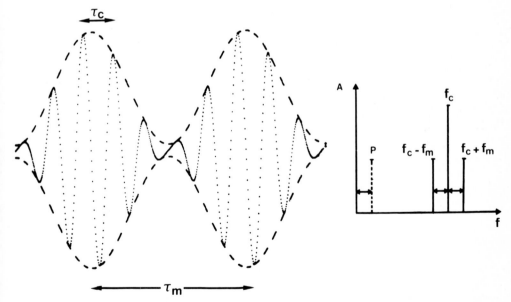

Fig. 1. Waveform and spectrum of an AM signal. The period of the envelope τ_m is the same as the period of the modulation frequency f_m. The period τ_c is the reverse of the carrier frequency $f_c = 1/\tau_c$. The spectrum consists of only 3 frequencies f_c, $f_c + f_m$, and $f_c - f_m$. The residual pitch P corresponds to a first approximation to the pitch of f_m. Note, that there is no frequency component f_m, which could be responsible for the perceived pitch.

AM signals with sinusoidal envelopes have only three frequency components: the carrier frequency f_c, $f_c + f_m$, and $f_c - f_m$, where f_m is the modulating frequency (Fig. 1). The residual pitch shifts as a function of f_c with a slope approximately given by the harmonic ratio $n_H = f_c/f_m$ (first effect of pitch shift). This effect may be explained by the assumption that the exact pitch value is defined by a subharmonic of the carrier. However, more thorough studies by Schouten et al. (1962) revealed small but systematic deviations of the slope to smaller values (second effect of pitch shift). These pitch effects could not be explained by the subharmonic hypothesis nor by other theories assuming, for example, that the exact pitch values are subharmonics of the lowest frequency component (Walliser 1969, Ritsma 1970). The temporal fine-structure mechanisms of de Boer (1956) was also found to be inadequate, since the details of the waveforms are strongly dependent on phase, and the residual pitch is not.

Only the pattern models by Terhardt, Goldstein, and Wightman (see above) were able to fit the data by assuming that every resolved frequency component including distortion products (Goldstein 1978) contribute to the residue. While the pitch estimates of Wightman's pattern transformation model without additional weighting functions fit the data qualitatively (Wightman 1973), the theories of Goldstein and Terhardt use special weighting functions to obtain quantitative correspondence. The essence of all three models is an internal (neuronal) representation of a more or less resolved frequency spectrum which would be necessary for the assumed auditory pattern analyzer to compute the residual pitch. As Wightman pointed out, his model is formally identical to an auto-correlation theory in the time domain, which implies that the same theoretical predictions of the data would be obtained by using a neuronal time code. However, no experimental evidence was found for neuronal correlation mechanisms in the time domain as required for example for the auto-correlation model of Licklider (1951). Hence, the advantage of the pattern models was that they could be based, in principle, on well investigated mechanisms, i.e. the frequency analysis of the cochlea, avoiding unknown temporal features of the auditory system. As Goldstein (1973) noted, "all current theories of periodicity pitch postulate some form of central processing based on peripheral frequency spectrum analysis".

As it turns out, these concepts are no longer tenable. One counter-evidence was already obtained by Ritsma (1962). By determing the "existence region" for residue signals with three frequency components he showed that just the harmonics best resolved by place coding, i.e. those with a harmonic ratio $n_H < 3$ do not or only weakly contribute to the residual pitch. On the other hand, for harmonics with $n_H > 4$ physiological results indicated that the frequency resolution of the cochlear filter is unsufficient for the postulated pattern analysis in contrast to the upper boundary for the residual pitch which is not below $n_H = 20$. This was recognized very soon, for example by Goldstein (1978): "neural time following, rather than details of the filter characteristics, is the decisive factor in determining the precision of aural frequency measurement". Consequently, a modern version of Goldstein's pattern model includes time coding of the frequency components (Srulovicz and Goldstein 1983) and thereby at the same time abandons the main theoretical advantage of the pattern models: to explain periodicity pitch on a purely spectral basis.

Moreover, by forward masking of periodic signals it was shown that not only the peripheral, but also the internal spectral analysis resolves only the first harmonics ($n_H \leq 4$) (Moore and Glasberg 1983).

These authors concluded that "the 'ripple' on the internal spectra of the maskers amounts to 3 dB or less for harmonics above the fifth". If this result holds, it is hard to see how frequency components can be used at all for a pattern analysis, since the lowest resolved harmonics are neglected and higher components - although clearly contributing to the perception of pitch - are only scarcely resolved, if at all.

THE PHYSIOLOGICAL APPROACH

It was logical to explain as much as possible by the relatively well investigated filter properties of the basilar membrane. Helmholtz's place principle turned out to be a powerful concept, especially with the modifications introduced by the pattern theories of pitch perception. However, neurophysiological investigations revealed more and more the restrictions for such theories.

The pure place principle implies that the frequency components of a signal are coded in neurons by spike rate versus place, which is identified in neurophysiological experiments by the frequency with the lowest threshold, the best frequency (BF). This code may be sufficient only for the first harmonics and only for low intensity levels. There is evidence that at medium and high levels (60 dB SPL) harmonics and especially formant frequencies are unsufficiently represented in the rate code (Sachs and Young 1979).

Since pitch (and by means of the formants also the vowels) are recognized even up to highest intensity levels, there must be another way for coding this auditory information. Single unit recordings have shown phase coupled responses at least up to 5 kHz (Kiang et al. 1965, Rose et al. 1967). In addition, Young and Sachs (1979) showed that, in contrast to place code, measures of frequency synchronous nerve activities, when plotted as a function of BF, preserve all necessary information about formants and periodicity at least up to 80 dB SPL. The obvious conclusion is that somewhere in the <u>central</u> auditory system there are neuronal mechanisms adequate for processing time information as encoded in the interspike intervals.

However, this does not imply the use of intervals corresponding to the envelope period for the purpose of pitch analysis. This possibility seems to be excluded by the interpretations given for Schouten's effects of pitch shift. On the other hand, provided the auditory system is actually able to analyse spike intervals, the demonstration of such intervals might be considered as an evidence for coding of periodicity pitch in the time domain (Evans 1983). At least higher harmonics with $n_H > 3$ interfere with each other on the basilar membrane - as Schouten (1940a) suggested - due to the limited resolution of the cochlear filter. These unresolved frequency components produce a vibration pattern of the membrane with an envelope period equal to the period of the fundamental. Evidence for an adequate interval coding was also demonstrated at the level of the auditory nerve (Evans 1983).

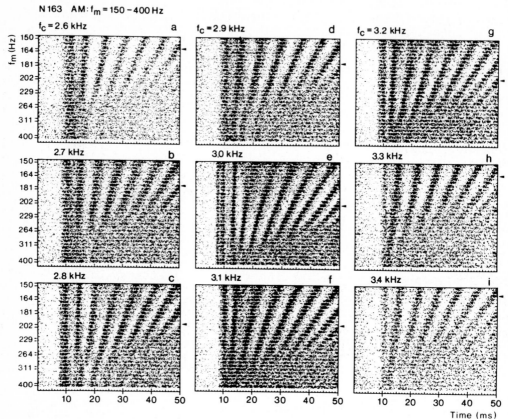

Fig. 2. Best modulation frequencies (BMF) vary as a function of the carrier frequency. Point-plot representation of the responses of a unit to AM signals in the midbrain (MLD) of the Guinea fowl. The modulation frequency was varied as indicated at the y-axes between 150 and 400 Hz. Nine carrier frequencies between 2.6 and 3.4 kHz were used. All signals started with zero crossings at zero on the time axis (100 repetitions). Only the first 50 ms is represented in the plots. The best modulation frequencies (BMFs) are indicated by arrows. They are characterized by the highest spike rate and by the best modulated responses in the point plots between 15 and 50 ms after stimulus onset. (Sound pressure level: 55 dB, 65 dB for (e); from Langner 1983).

Binaural psychophysical pitch effects (see below) require periodicity analysis at a higher level of the auditory system where information from both ears converge. Consequently, it should be expected to find neuronal representation of the signal envelope at the level of the midbrain. This was demonstrated in the MLD nucleus of the Guinea fowl for stimulus parameters of AM signals, which were typical for the residual effect, i.e. for envelope frequencies below 1200 Hz and carrier frequencies below 5 kHz (Langner 1978, 1981, 1983a). The results obtained in the midbrain of the bird suggest that the auditory system performs a cross-correlation analysis between spike activities coupled on the one hand to the envelope and on the other hand to the carrier of the signal (Fig. 2).

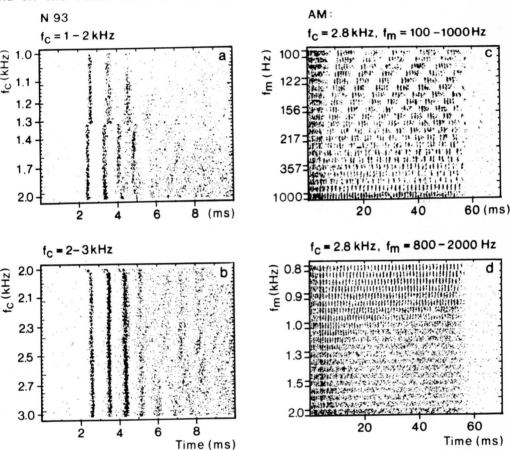

Fig. 3. *Intrinsic oscillations triggered by amplitude fluctuations. a, b: Responses of a MLD-unit in the Guinea fowl to 4l unmodulated tone-bursts (50 repetitions). Each tone started with a zero crossing at zero on the time axis. The unit (BF = 2.3 kHz) shows oscillatory responses with periods of about 2·0.4 ms triggered by the onsets. Note, the influence of frequencies on the intrinsic oscillations varying in response strength, number and exact values of the periods. c, d: Responses to AM signals with constant f_c and varying f_m. Not only the stimulus onsets, but every AM cycle, triggered an intrinsic oscillation which changed into a coupled oscillation for f_m about 800 Hz. (From Langner 1983).*

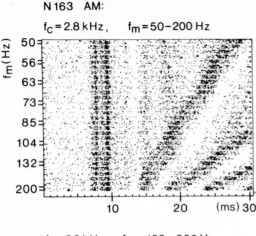

N 163 AM:

$f_C = 2.8$ kHz, $f_m = 50-200$ Hz

f_m (Hz)

50
56
63
73
85
104
132
200

10   20   (ms) 30

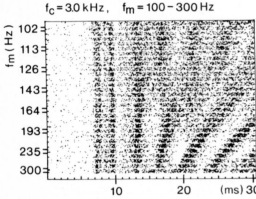

$f_C = 3.0$ kHz, $f_m = 100-300$ Hz

f_m (Hz)

102
113
126
143
164
193
235
300

10   20   (ms) 30

Fig. 4. Coincidence effects. Responses of a midbrain unit in the Guinea fowl to AM signals with constant f_c for each point plot. f_m varying as indicated along the y-axes (100 repetitions, 65 dB SPL). Only the first 30 ms after stimulus onset were plotted. The responses were restricted to the range of the BF = 2.9 kHz. In the upper part of the plots several oscillations (vertical contours) in response to the first modulation at the onset of the signal are visible. The coincidences of these oscillations with the responses to the following cycles (sloping contours) resulted in phase shifts: without the coincidence effect the response contours would be expected to be straight. In the areas of intersection of the vertical with the sloping contours the response strength was enhanced, in between these areas along the sloping contours the spike pattern is blurred.

Cross-correlations are used in system analysis to measure the similarity of two signals. One signal is delayed and then multiplied with the second signal. The result is integrated over a relevant time window. Common frequency components or harmonic relations within the compared signals are indicated by maxima of the resulting cross-correlation function. In order to demonstrate cross-correlation in the central nervous system it is therefore necessary to demonstrate delay mechanisms and multiplication, i.e. coincidence mechanisms. In the auditory system the delays may be provided by intrinsic oscillations as described for the midbrain of the Guinea fowl (Figs. 3, 4, and 5). They may be triggered by envelope fluctuations and last for several milliseconds. Every spike of the oscillation represents a delayed response to a certain trigger point near the zero crossing of the modulating waveform. In AM signals every cycle triggers such an oscillation (Fig. 3). The periods of the intrinsic oscillations have preferred values of 0.8 ms or greater integral multiples of 0.4 ms, which may be explained by a neuronal model using synaptic delays of 0.4 ms (Fig. 5). Synaptic delays of this length (0.4-0.6 ms) were proposed for example by Pfeiffer (1966a) for the cochlear nucleus of the cat. Chopper responses as first described by Pfeiffer (1966b) for the cochlear nucleus of the cat and characterized by regular spike intervals may well be the origin of the oscillations observed at the level of the midbrain.

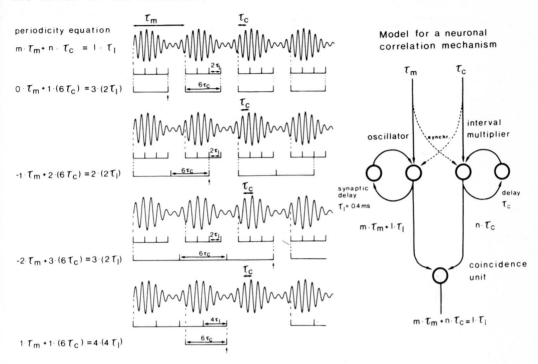

Fig. 5. *Periodicity analysis in the time domain. Left: Each AM cycle triggers intrinsic oscillations (upper "spike trains") with periods of 0.8 ms (resp. 1.6 ms at the bottom). An interval multiplier generates intervals 6 τ_c (lower "spike trains") synchronized at the beginning with the oscillations. Four different τ_c's produce four different coincidence effects (marked by arrows) expressed mathematically by the periodicity equations. Right: The 3 main components of the model are an oscillator, a multiplier, and a coincidence unit. The spike trains from the oscillator and the multiplier are triggered together by the AM-cycles. The excitation in the oscillator circles with a delay of 0.4 ms at each synapse resulting in a period of 0.8 ms when only two units are involved. The delay at a multiplier unit is τ_c (or a multiple) if the two units activate each other below threshold and have additional inputs coupled to the carrier. The coincidence unit (conceptualized at the level of the midbrain) is activated when a spike from the oscillator coincides with a spike from the multiplier. (From Langner 1983).*

Multiplication may be obtained in neuronal networks when two input spikes coincide at the level of a postsynaptic unit. Evidence for such coincidence effects due to responses to the envelope of AM signals on one side and to the carrier on the other side comes mainly from three observations:
1. The best modulation frequency (BMF) of a given unit varies with

the carrier frequency of the AM signal (Fig. 2).

2. The response at a certain point of time is strong when delayed spikes, due to the intrinsic oscillations, coincide with undelayed responses to the AM cycles (Fig. 4).

3. The phase delay of the response may be measured in relation to the modulation or the carrier frequency. In the range of frequencies giving rise to a coincidence, changes of the phase delay were observed (Fig. 4). Quantitatively these changes could be described as linear functions of the period of the varying frequency, i.e. either of the modulation or of the carrier. In this coincidence range the relation between BMFs and carriers may be described by the periodicity equation:

$$m \cdot \tau_m + n \cdot \tau_c = l \cdot \tau_1,$$

where τ_1 = 0.4 ms and m, n, and l are small integers (m mostly 1 or 2; l>1). In order to explain why the value n seemed to be restricted to certain integer values for a given unit, additional mechanisms had to be assumed. According to the volley principle (Wever 1949) carrier frequencies may be coded in spike intervals up to about 5 kHz, presumably the upper limit of phase coupling. At the input of the coincidence unit these intervals seem to be reduced to $n \cdot \tau_c$ (in the order of τ_m). In more sophisticated models it may be sufficient to attribute these intervals to a higher probability than other intervals, as already found at the level of the auditory nerve (Evans 1978). A possible theoretical explanation of these results is given in Fig. 5.

The results, obtained in the avian midbrain, including the same time constant of 0.4 ms, were confirmed by recordings of AM responses in 760 units in the midbrain (inferior colliculus) of the cat (Langner and Schreiner, in prep.). Signal envelopes were found to be represented up to about 1000 Hz in the neuronal time patterns, though BMFs above 300 Hz were demonstrated only by multi-unit recordings. The upper limit for envelope coding is expected at 1250 Hz corresponding to the shortest possible intrinsic oscillation period of 0.8 ms. Another boundary was found for harmonic ratios $n_H = f_c/f_m = 4$, i.e. only few units revealed BMFs corresponding to a smaller ratio of carrier and modulation frequency. This may be explained by the better resolution of lower harmonics with $n_H < 4$ limiting the auditory coding capability for the envelope. On the other hand, phase coupling is best at low frequencies (Johnson 1980). Consequently, cross-correlation between carrier and envelope will reveal best results for the lowest harmonics of a signal exceeding $n_H = 3$. This may explain the "dominance concept" of the residual pitch favouring the harmonics around $n_H = 4$ (Ritsma 1970).

Another evidence for the importance of periodicity coding in the auditory system comes from the demonstration of a topographic organization of BMFs within the central nucleus of the cat's inferior colliculus. BMFs were mapped on isofrequency planes of the tonotopic organization in the nucleus. Similar BMFs are located in each plane on concentric lines around a maximal BMF (Schreiner und Langner in prep.).

PERIODICITY PITCH BASED ON TIME CODING

It appears that periodicity coding in birds and cats is analyzed by neuronal correlation mechanisms in the time domain, and that similar mechanisms in humans may be involved in pitch perception. The pitch of a residue perceived by a human subject may be measured by the period τ_p of a pure tone, which give rise to the same pitch.

From the cross-correlation described above and expressed by the periodicity equation, the envelope period may be obtained as a linear function of the carrier period τ_c. Consequently, for the period τ_p the following relation should be expected:

$$\tau_p = n \cdot \tau_c + b.$$

While $n \cdot \tau_c$ stands for intervals which are multiples of the carrier period τ_c, the value b expresses the correlational delay introduced by intrinsic oscillations. Since n is an integer this pitch equation indicates that the residue of AM signals is related to a sub-harmonic of the carrier frequency, which could explain the first effect of pitch shift. However, the intercept b is not zero and, consequently, τ_p is not simply harmonically related to τ_c (i.e. τ_p/τ_c non-integer). This deviation from harmony may explain the second effect of pitch shift with nonharmonic AM signals.

As well known, pitch perception is highly ambiguous (de Boer 1976). This was confirmed by results obtained with 5 human subjects (Langner 1983b). For a certain value of τ_c, i.e. for a certain carrier frequency, several values for n and b were obtained, depending on the subject and the experimental conditions. According to the pitch equation given above the results in humans suggested that the preferred values for n are integer near the harmonic relation $n_H = f_c/f_m$, and the preferred values for b are multiples of 0.4 ms. These results suggest that the pitch perception in humans may be based on similar correlation mechanisms, including the same time constant as the periodicity analysis in the experimental animals.

Other evidence was the observation that the residue pitch did not always increase linearly as predicted by the pitch equation. In some subjects and only when the carrier frequency was increased in small steps (10 Hz), the pitch curves revealed steps whenever $n_H \cdot \tau_c$ reached multiples of 0.4 ms: $l \cdot 0.4$ ms. For a certain range of τ_c the values of the pitch period were observed to be also the same multiple of 0.4 ms: $\tau_p = l \cdot 0.4$ ms $\approx n_H \cdot \tau_c$ (Langner 1981, 1983b).

As the pitch perception indicates, the auditory system may determine the value of τ_p by the cross correlation. Consequently, all parameters on the right side of the pitch equation, i.e. n, τ_c, and b must be available and orderly represented at the level of the analysis. The parameters n and b should be topographically represented by equivalent neuronal properties. In contrast, the carrier period, as a signal parameter, must be defined either by the correlation mechanisms (perhaps for low frequencies) or by means of the place principle (probably for high frequencies). As results from Young and Sachs (1979) indicated: some form of place information is available provided that time coding is utilized. Note that these mechanisms will fail above 5 kHz where phase coupling deteriorates, explaining the upper limit of the "existence region" for the residue. The difference limen for periodicity discrimination above 5 kHz is predicted by this theory to be in the order of 0.4 ms, because the _cross_-correlation can only be replaced by an _auto_-correlation using again multiples of 0.4 ms as delays.

In the classic experiments of Schouten et al (1962) the residual pitch of nonharmonic AM signals (f_c/f_m non-integer) were compared with the pitch of harmonic AM signals ($f_c/f_m = n_H$ integer). Under such carefully selected experimental conditions the attention of the subjects might be focussed on the strongest coincidence effects reducing pitch ambiguity. Many units in the midbrain of both cats and birds were found to have an intrinsic oscillation period of 0.8 ms, which is also the shortest possible. From the periodicity equation the prominent coincidence effect is given by: $\tau_m + 0.8$ ms $= n \cdot \tau_c$. On the other hand, the pitch experiments of Schouten et al were concentrated on carrier frequencies in the range of the harmonic point, where $\tau_m = n_H \cdot \tau_c$. Consequently, the expected _mean_ value a for n is given by: $a = \tau_m/\tau_c + 0.8$ ms$/\tau_c = n_H + 0.8$ ms$/(\tau_m/n_H) = n_H(1 + 0.8/\tau_m) = 1.16 \, n_H$, when $\tau_m = 5$ ms as in these experiments (Fig. 6).

There is another classic experiment indicating that pitch must be analyzed in the time domain. If wideband Gaussian noise from one signal source is presented binaurally, an interaural delay τ_d will be perceived as a pitch corresponding to the pitch of a tone with a period $\tau_p = \tau_d$ (Bilsen and Goldstein 1974). In this case pitch must be exclusively coded in the time differences between phase coupled activities in both ears, i.e. in the corresponding parts of the central auditory system. As each event (i.e. amplitude fluctuation) in a certain filter channel of one ear only is delayed for the corresponding channel in the other ear, the delay may be exactly measured by the central coincidence mechanisms, as shown by the results. Introducing an additional phase shift of 180° for all frequencies will abolish all coincidences obtained previously. Moreover, the delays will now be frequency dependent and consequently different for each frequency channel. However, coincidences evoked by corresponding events in both ears will still be possible, of course in all those frequency channels where the additional delays may be compensated by the triggered intrinsic oscillations. The compensation by oscillations, triggered in the same ear which receives the delayed signal, will result in slightly over-estimated delay values. However, at the same time the compensation may take place using the oscillations triggered in the other ear. In this case the delay value will be under-estimated. Since the prominent interval of these oscillations will be again 0.8 ms, the theory is completely in line with the pitch values observed in this experiment: $\tau_d + 0.8$ ms and $\tau_d - 0.8$ ms.

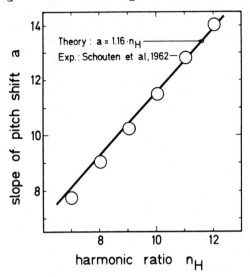

Fig. 6. Predictions of pitch measurements. Pitch values for AM signals were obtained by Schouten et al (1962) with f_c varying around harmonic relations $f_c = n_H \cdot f_m$. Their results were replotted with τ_p (period of the sine wave with the same pitch as the AM signal) as a function of the carrier period instead of the carrier frequency. The data points in the plot are the slopes of the resulting pitch curves. The straight line derived from the theoretical prediction for the pitch period $\tau_p = n \cdot \tau_c - 0.8$ ms and the fact that for $\tau_m = n_H \cdot \tau_c$: $\tau_p = \tau_m$ (= 5 ms for these experiments). Note, that the neuronal correlation theory predicts the slope of the pitch curves and the pitch values without the use of any additional fitting parameters.

CONCLUSIONS

1. Periodicity analysis is an important task of the auditory system, as demonstrated by the topological organization of best modulation frequencies on concentric lines on the isofrequency laminae in the midbrain of the cat.
2. As predicted by Schouten (1940a), the auditory system may make use of the envelope of periodic signals to define their pitch values. This is in contrast to the pattern models which assume that pitch is based on resolved frequency components. Nevertheless, these models had some success in explaining various pitch effects, which may be explained by their formal correspondence to a correlation analysis in the time domain.
3. Although the envelope may be coded by interspike intervals, the envelope period apparently does not define the pitch of the residue directly, but only after a cross-correlation with intervals coding the carrier. Making use of Wever's volley principle (1949), this

mechanism is limited by the upper boundary for phase coupling, probably at 5 kHz.

4. Delays necessary for the correlation are provided by intrinsic oscillations triggered by envelope fluctuations. The periods of these intrinsic oscillations are 0.8 ms or greater multiples of 0.4 ms.

5. The present pitch theory is in line with basic neurophysiological and psychophysical results like phase coupling properties and limited frequency resolution of the auditory analysis. It may also explain important results of pitch experiments like the first and second effect of pitch shift, including exact predictions of the pitch values, the dominance region and binaural pitch effects.

ACKNOWLEDGEMENT

This study was supported by the Deutsche Forschungsgemeinschaft (SFB 45). The results in cats were obtained together with Dr. C. Schreiner at the Coleman Lab. of the University of California in San Francisco. I wish to thank Prof. Dr. M. Merzenich, Prof. Dr. H. Scheich and Mr. M. Camargo for support of experiments.

REFERENCES

Bilsen FA, Goldstein JL (1974) Pitch of dichotically delayed noise and its possible spectral basis. J Acoust Soc Amer 55:292-296

Boer E de (1956) Pitch of inharmonic signals. Nature 178:535-536

Boer E de (1976) On the "residue" and auditory pitch perception. In: Keidel WP and Neff WD (eds) Handbook of sensory physiology. V/3, Springer, Berlin, pp 479-583

Evans EF (1978) Place and time coding of frequency in the peripheral auditory system: some physiological pros and cons. Audiology 17:369-420

Evans EF (1983) Pitch and cochlear nerve fibre temporal discharge patterns. In: Klinke R and Hartmann R (eds) Hearing - physiological bases and psychophysics. Springer, Berlin, pp 140-145

Goldstein JL (1973) An optimum processor theory for the central formation of the pitch of complex tones. J Acoust Soc Amer 54:1496-1516

Goldstein JL (1978) Mechanisms of signal analysis and pattern perception in periodicity pitch. Audiology 17:421-445

Helmholtz H von (1862) Die Lehre von den Tonempfindungen als physiologische Grundlage für die Theorie der Musik. Vieweg, Braunschweig

Johnson DH (1980) The relationship between spike rate and synchrony in responses of auditory-nerve fibers to single tones. J Acoust Soc Amer 68:1115-1122

Kiang NYS, Watanabe T, Thomas ED, Clark LF (1965) Discharge patterns of single fibers in the cat's auditory nerve. MIT-Press, Cambridge, Mass.

Langner G (1978) The periodicity matrix. A correlation model for central auditory frequency analysis. Verh Dtsch Zool Ges., p 194

Langner G (1981) Neuronal mechanisms of pitch analysis in the time domain. Exp Brain Res 44:450-454

Langner G (1983a) Evidence for neuronal periodicity detection in the auditory system of the Guinea fowl: implications for pitch analysis in the time domain. Exp Brain Res 52:333-355

Langner G (1983b) Pitch of AM-signals: Evidence for a correlation analysis in the human auditory system. Soc Neuroscience, 13th annual meeting, Boston, 193.3

Langner G, Schreiner C (1985) Periodicity coding in the inferior colliculus of the cat: I. Neuronal mechanisms (in prep.)

Licklider JCR (1951) A duplex theory of pitch perception. Experientia 7/4:128-134

Moore BCJ, Glasberg BR (1983) Forward masking patterns for harmonic complex tones. J Acoust Soc Amer 73:1682-1685

Ohm GS (1843) Über die Definition des Tones, nebst daran geknüpfter Theorie der Sirene und ähnlicher tonbildender Vorrichtungen. Ann Phys Chem 59:513-565

Pfeiffer RR (1966a) Anteroventral cochlear nucleus: wave forms of extracellularly recorded spike potentials. Science 154:667-668

Pfeiffer RR (1966b) Classification of response patterns of spike discharges for units in the cochlear nucleus: tone burst stimulation. Exp Brain Res 1:220-235

Ritsma RJ (1962) Existence region of the tonal residue. J Acoust Soc Amer 34:1224-1229

Ritsma RJ (1970) Periodicity detection. In: Frequency analysis and periodicity detection in hearing. (Plomp R and Smoorenburg GF, eds), Sijthoff, Leiden, pp 250-263

Rose JE, Brugge JF, Anderson DJ, Hind JE (1967) Phase-locked response to low-frequency tones in single auditory nerve fibers of the squirrel monkey. J Neurophysiol 27:768-787

Sachs MB, Young ED (1979) Encoding of steady-state vowels in the auditory nerve: representation in terms of discharge rate. J Acoust Soc Amer 66:470-479

Schouten JF (1940a) The residue, a new component in subjective sound analysis. Proc Kon Acad Wetensch 43:356-365

Schouten JF (1940b) De toonhoogtegewaarwording. Philips Technisch Tijdschr 5:298-306

Schouten JF, Ritsma RJ, Cardozo BL (1962) Pitch of the residue. J Acoust Soc Amer 34:1418-1424

Seebeck A (1841) Beobachtungen über einige Bedingungen der Entstehung von Tönen. Ann Phys Chem 53:417-436

Schreiner C, Langner G (1985) Periodicity coding in the inferior colliculus of the cat: II. Topographical organization (in prep.)

Srulovicz P, Goldstein JL (1983) A central spectrum model: a synthesis of auditory-nerve timing and place cues in monaural communication of frequency spectrum. J Acoust Soc Amer 73:1266-1276

Terhardt E (1972) Zur Tonhöhenwahrnehmung von Klängen. II. Ein Funktionsschema. Acustica 26:187-199

Walliser K (1969) Über ein Funktionsschema für die Bildung der Periodentonhöhe aus dem Schallreiz. Kybernetik 6:65-72

Wever EG (1949) Theory of hearing. Wiley, New York

Wightman FL (1973) The pattern-transformation model of pitch. J Acoust Soc Amer 54:407-416

Young ED, Sachs MB (1979) Representation of steady-state vowels in the temporal aspects of the discharge patterns of populations of auditory nerve fibers. J Acoust Soc Amer 66:1381-1403

Temporal Factors in Psychoacoustics

David M. Green

Laboratory of Psychophysics
Harvard University, 33 Kirkland Street, Cambridge,
Massachusetts 02138, USA

INTRODUCTION

The goal of psychoacoustics is to provide a mechanistic description of the listener's auditory processes. Because any such process operates in time, it is clear that temporal factors must play a role in their description. Some simple, first-order questions might concern how quickly a certain mechanism can operate, or, on the other hand, what is the longest time over which it can integrate or combine acoustic information. Other sorts of information concerning the temporal scale are relevant in other investigations. Because of the diverse nature of the different studies in this field, there is hardly one answer to the variety of questions that one can ask about temporal phenomena.

Thus, the viewpoint assumed throughout this discussion is that temporal phenomena manifest themselves in different ways in different psychophysical experiments. Statements about temporal mechanisms should be specific about the task of the observer in the experiment. Detection of a faint sound occurring in quiet may produce quite different temporal parameters from those found when measuring the detection of a short increment in continuous noise. General statements about temporal parameters are probably a thing of the past. In my view, it is similar to discussing the speed of a chemical process. No broad statement about this class of phenomena is likely to have much generality. They will obviously differ, depending on the chemical and the process.

What is even more insidious in my specialty is that the difference can also depend on the specific aims of the listener in the particular psychoacoustic task. Detecting a small increment in a noise burst and trying to detect a brief silent interval may produce very different estimates of the ear's temporal parameters. Sometimes we are fortunate and can try to account for the discrepancies as different aspects or expressions of the same basic mechanism. Certainly, parsimony recommends this strategy as a prudent first step. But often the differences are so great that we are forced to conclude that there are, in fact, two very different mechanisms available to the listener and that different tasks cause the listener to use one or the other process. While such interpretations are still not popular in some quarters, they are gaining more general acceptance, especially from those who view a great deal of sensory psychology as "information processing". Such a viewpoint almost always implies that the explanations will be complex and depend on a variety of non-sensory factors. But, for those adopting such a perspective, this level of complexity is no more than appropriate, given the diversity of the organism's behavior and the enormous number of computational elements present in the human brain.

While the information processing view is hardly my main outlook in psychoacoustics, it clearly has some validity and consequently makes my job of writing about temporal mechanisms inferred from psychoacoustic experiments difficult. Given this background, I will review several specific areas where extensive work on the temporal mechanism has been pursued and summarize the knowledge we have gained in these specific areas. I will also try to list the different types of experimental procedures that have been used in an attempt to gain insight into the temporal aspects of the mechanism under study. In many ways, these different methods are as interesting and potentially useful as anything of substance we have yet discovered.

I will distinguish between two broad classes of temporal phenomena. These two classes represent polar opposites along the temporal continuum and both have received considerable attention in our contemporary research. One class is temporal integration. How long can we accumulate or integrate acoustic information? Generally speaking, the type of task that addresses this kind of question involves the detection of barely audible signals. Longer integration times mean better detection of the weak acoustic signal and it is important to establish the maximum integration time that can be used by the observer in this type of task. The almost universal approach to this problem is to study the time-intensity trade. If we increase the duration of the signal and the observer is still integrating acoustic information, then one should be able to trade time for intensity and still maintain the same level of detection performance. When integration fails, the signal's detectability is determined solely by intensity. Thus, the time-intensity trade amounts to measuring a function (signal intensity versus signal duration for a constant level of detectability) and determining the break point in this function.

At the other extreme of the temporal continuum is what I have called temporal acuity. The interest here is how fast can the auditory system work. Given brief acoustic events or small differences in the starting time of one or more signals, can the observer discriminate such small differences in time? Because of the inertia in the processing of acoustic information, there must be some minimal temporal differences that are indistinguishable. What are these minimal times? This is an area of considerable subtlety, since changes in temporal events produce changes in other aspects of the acoustic stimulus that may provide the basis of the discrimination. These artifacts obscure simple answers to our original question and greatly complicate the interpretation of experimental data in this area of research. Regrettably, there is also a sizeable range for the estimated values. This fact further complicates our understanding in this area.

Let us begin this review by considering the integration process and what we know about this temporal parameter.

INTEGRATION TIME IN DETECTION TASKS

TIME-INTENSITY TRADE

One of the oldest and most popular methods for studying the temporal properties of any sensory system is the time-intensity trade. In its simplest form, this paradigm involves measuring the detectability of the signal as a function of its duration. One then constructs a plot of signal intensity versus signal duration for a constant level of detectability. The results are usually of the following sort. At the shortest durations, one usually finds that increasing the signal's duration makes it possible to detect the signal at a lower intensity value. At longer durations, however, the trading ratio diminishes and eventually the signal's intensity threshold becomes nearly independent of duration. This value, beyond which duration has little or no effect, is the "critical duration" and is taken as an estimate of the integration time of the device.

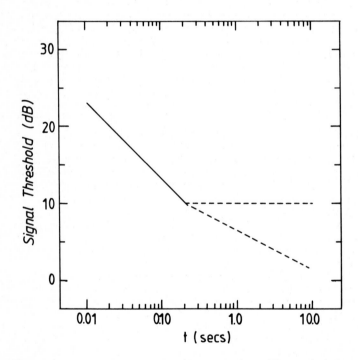

Fig. 1. Typical time-intensity trading results. The solid line shows the improvement in threshold as a function of signal duration. The dotted lines show two alternatives. The line with zero slope shows no improvement in threshold with duration. The other dotted line shows improvement at the rate of 1.5 dB per double of duration.

There are several methodological points that should be noted about this estimate. First, this integration time is a break in the function relating signal intensity and duration, as indicated in Fig. 1. Ideally, one would hope to find that the function, on either side of the breakpoint, would fall along roughly straight lines. In that case, the intersections of these two lines would be the desired parameter estimate. For time-intensity trades, for example, one often finds an energy rule (the product of time and intensity is constant) to the left of the breakpoint, and constant intensity to the right of the breakpoint. In this case, the breakpoint is the intersection of two lines meeting at a 45 degree angle. There are two problems with estimating the value at which this intersection occurs.

First, it is difficult to get as many points on either line segment as one would wish. Obviously, a larger number of points are desirable because more points increase the definitions of each line segment and hence clarify the point of intersection. But the most useful data points are those at some distance from the breakpoint. Only these points are unambiguously connected to one or the other line segment. Also, for very brief sinusoidal pulses, we know the energy rule can only operate over a restricted range. Short duration sinusoids have energy splattered over a wide frequency region. This fact alone is likely to cause the simple energy rule to fail. In practice then, the line to the left of the breakpoint is often measurable only over a single decade of duration or less. Second, it is tedious to obtain measurements at very long durations, so that values spanning a full log unit at long durations are also rare.

More fundamental, however, is the possibility that the assumption that detection is independent of duration for the long duration signals may be wrong. A simple Gedanken experiment should convince us that the intuitive expectation of a flat line at very long durations may be compromised. Consider two very long duration signals, say a five-minute and a ten-minute signal, of some constant intensity level. Now suppose the observer listens to the sound at regular intervals, say every 10 seconds. The observer simply records whether the signal appears to be absent or present. We may assume that for any intensity level of the signal the probability of deciding that the signal was present is some fixed probability p. The number of "signal present" decisions divided by the total number of these observations is an estimate of that probability p. The stability of that estimate improves as the number of decisions increases. Hence, a ten-minute signal should be more detectable than a five-minute signal based only on these simple binomial considerations.

What we have been calculating is essentially the d' values for the two signal durations. For the long duration signals, this simple binomial process suggests that d' will increase with the square root of duration. For many detection tasks, we find that d' is nearly proportional to intensity. Hence, in these situations, we would find that the threshold of the signal would not be independent of the signal duration, but would decline as the square root of duration. Estimating the intersection of our two line segments is more difficult, since the change in slope of the two straight line segments is now only 22.5 degrees.

Obviously, we seldom have the patience to explore very long duration conditions with care. Such exploration is especially tedious if we use one of the forced-choice techniques, since the length of a typical trial cycle increases with the number of alternatives used in the test. One can see in the existing data, however, a small but consistent tendency for the signal threshold to improve slightly with increasing duration. These facts increase our uncertainty as to the estimate of the integration time. Nonetheless, many estimates of these critical durations have been carried out and, although some scatter is evident, the general agreement is gratifying.

As a single summary number, it seems clear that a 200 msec critical duration comes close to all existing data. There is good evidence that low frequency sinusoids have somewhat larger critical durations than higher frequency sinusoids. Plomp and Bouman (1959), for example, estimate the time-constant to be 375 msec at 200 Hz and about 150 msec at 8000 Hz. Watson and Gengel (1969) have also confirmed this trend. Their paper provides the best modern review of this topic as well as some interesting experimental data. Zwislocki (1960) has developed a theory of temporal summation that is based largely on a simple exponential model. He shows how this model, using a 200 msec time-constant, provides a good account of a variety of data related to the detection of brief acoustic pulses as well as sinusoids. The same exponential model is the basis of theories by Munson (1947) and Plomp and Bouman (1959). The existing experimental data are largely in agreement, and it would require a massive amount of new data to convince us that the critical duration for such experiments is far from the 200 msec value.

While the data on judgments of the loudness of sounds generally parallel those obtained from detection tasks, there are some notable and sizeable discrepancies. Many studies find the critical duration of about 200 msec (the range is about 100 to 500 msec), but occasional estimates are in the tens of milliseconds range (Small, Brandt and Cox 1962). A complete review of the various studies may be found in Scharf (1978).

This completes the review of this end of the temporal continuum. Let us now turn to the opposite end of the time scale and studies that estimate how fast the auditory system can operate.

AUDITORY TEMPORAL ACUITY

So far, we have discussed studies of integration processes. Such studies measure the longest or maximum time over which the ear can combine or average acoustic information. But what of the other side of the temporal dimension? What is the fastest the ear can react, and how quickly can it process a stream of acoustic events so as to distinguish among them? I have called the measurement of this aspect of auditory temporal processing auditory temporal acuity (Green, 1971). The analog with the spatial acuity in vision is apt because in test of visual acuity we wish to know the minimal discriminable separation between two objects whereas in hearing we wish to know the minimal temporal separation. What is the smallest time that still permits us to determine that two events occurred and to discriminate the order in which they occurred. While, in principle the measurement of this minimal time is simple, in practice it is more complicated. The major problem is that the discrimination among acoustic events can frequently be accomplished on the basis of cues that have little to do with temporal resolution. Changes in the relative position of events in time almost inevitably produce changes in the long-term auditory power spectrum. Thus, for example, discrimination of one versus two acoustic events may be possible because of the presence or absence of energy at some particular frequency region. The detection of this energy and hence the discrimination of one versus two events can be accomplished on the basis of long-term integration of energy at that region (Leshowitz 1971). Such artifacts would completely invalidate any inferences one might wish to draw concerning the speed of auditory processing.

For these reasons, I have advocated that discrimination tests of this type be conducted only when the power spectra of the two stimuli are identical. Clearly, if both power and phase spectra are identical, then there is no difference in the waveform of the two stimuli and absolutely no physical basis for the discrimination test. But, if the two stimuli have identical power spectra, then the only possible difference in the stimuli must be the phase spectra. This does not mean that the basis of the discrimination must be discriminatory of changes in the phase spectra. Phase discrimination might be the basis for discrimination, but it is not necessarily the relevant one. This is true, since we can only equate the power spectra of two different waveforms over some definite interval of time (0,T). Computation of the power spectrum over a shorter interval of time will, by necessity, reveal differences in the two sounds unless they are, in fact, identical.

The rigorous experimental approach to the question of temporal acuity, then, involves presenting two sounds that are non-zero only in an interval (0,T). Outside that interval they are zero, and hence, the basis of discrimination between two waveforms is some computation made on the waveshape within the interval. That computation might be comparison of short-term power spectra, computation of effective duration, or a computation of the phase spectra. Whatever it is, the computation must be carried out within the interval (0,T), because only in that interval do the waveforms differ.

Now suppose we reduce that interval, making T shorter and shorter. Inevitably, the discrimination will fail. At that duration, we have an estimate of the temporal acuity of the system. Note: this experiment will not reveal, by itself, the basis of the discrimination at longer intervals. The critical features for the discrimination of the difference in the waveform might be a difference in phase, in effective duration, or in the short-term power spectra. Other experiments must be conducted to isolate the critical cue(s). What we have achieved is an estimate of the minimal time over which differences in the waveforms can be distinguished, independent of how they are distinguished.

Experimentally, generating waveforms with identical power spectra requires great care. If discrete spectra are produced, one must search the entire spectra diligently to be sure that spurious components do not arise as artifacts. Energy down 60 dB or more from the main components of the spectra may be clearly audible and provide a means of discriminating between the two stimuli. If continuous spectra are used, then care must be taken to avoid small perturbations in the shape of the spectra, since ripples as small as 0.25 dB may be audible. Two general techniques have been used: 1) time reversed waveforms, and 2) Huffman sequences. We will discuss them in reverse order, since there is more controversy about the results obtained with time reversed waveforms.

Huffman sequences are a digital signal technique developed for radar application (Huffman 1962). An entire class of waveforms can be produced, all of which are transient waveforms that are non-zero in an interval (0,T), with nearly flat power spectra. Members of the same class have identical power spectra but can differ in their phase spectra in a number of ways. Some intuition about these differences can be gained by viewing them as signals produced by digital all-pass filters. Different frequencies and bandwidths can be selected for the all-pass filters. Each filter then delays energy in the spectra around the center frequency of the filter by an amount and over a range of frequencies determined by the bandwidth of the filter. Patterson and Green (1970) were the first to study such signals in psychoacoustic tests. They generated pairs of Huffman signals that differed in some way and measured the ability to distinguish between the pairs as a function of the total duration of the signal, T. In effect, a psychometric function could be measured with discrimination varying between 50 and 100% as a function of the duration of the signals.

Selecting two signals with all-pass filters at 800 and 1600 Hz, they found that discrimination of the two signals could be accomplished when the duration exceeded about 2.5 msec. Other parameters such as the bandwidth of the all-pass filter, or the difference in frequency separation of the center frequencies of the two all-pass filters were also investigated. Manipulation of such variables changed the ability to discriminate among the signals in sensible ways. Green (1971) has presented a review of many of these studies. His summary for the temporal acuity number is 2 msec.

A second, and somewhat simpler, way to study temporal acuity is to use time-reversed stimuli. Ronken (1970) was among the first to pursue this approach. He used two clicks, one more intense than the other. At long durations between the two clicks one hears either "loud-soft" or "soft-loud", depending on the order in which the clicks are played. As the duration between the two clicks is diminished, discrimination of the order becomes more difficult. The phenomenal impression also changes, and one listens for slight differences in the quality of the sound, something like the difference in sound between "tick" and "tock".

Discrimination of the temporal order of two sounds can occur only if some other difference exists between the two sounds. In Ronken's experiment, this difference was the intensity of the two clicks. Clearly, if the two clicks were nearly equal in intensity, then the discrimination of their temporal order would be impossible. Also, if the less intense click is inaudible, then discrimination is bound to fail. Ronken found that discrimination of temporal order could be made at temporal separations of 1 to 2 msec if the intensity differences were reasonable (3-10 dB). Recent studies have repeated and extended Ronken's original work. Feth and O'Malley (1977) and Resnick and Feth (1975) have found that temporal acuity for clicks of different intensity is in the 1 to 2 msec range. They have argued that changes in the quality of the sound produced by forward masking produce this discrimination. In effect, the more intense click preceding the weaker one produces more masking than the reverse order. Ronken (1970), however, explicitly considers that hypothesis and rejects it, see his Appendix B. More troubling than these arguments about the basis for the temporal discrimination are the recent experimental results presented by Henning and Gaskell (1981). They used the two-click procedure and show good discrimination occurring in the less than 0.5 msec range for two of three observers. One observer, Dr. Gaskell, achieves better than 90% discrimination listening to two clicks separated by 250 microseconds. The only difference between this experiment and the several others is that the click duration in the Henning and Gaskell study was only 20 microseconds. All the other studies used much longer click durations, about 100 microseconds. There are probably no important differences in the spectra of 20 and 100 microsecond clicks, given the frequency response of most headphones. The shorter clicks however are probably presented at somewhat lower intensity levels than are the 100 microsecond pulses. Apparently Henning and Gaskell's level are about 10 to 20 dB lower in sensation level than most of the previous studies. Recently, in our laboratory, we listened informally to the 250 microsecond separation using 20 microsecond clicks and could not achieve better than chance performance.

Henning and Gaskell, in a theoretical analysis of their results, suggest that the discrimination of the temporal order of the clicks may be based on differences in the rms duration of the click pairs. They suggest that the critical experiment is to equate both the power spectra and the rms durations of the two stimuli, but could not accomplish this objective with their equipment.

In summary, studies of temporal acuity have found that discrimination is possible with times as short as 1 to 2 msec, ignoring the singular point at 250 microseconds. The basis of this discrimination is still under dispute but forward masking or difference in the effective duration of the stimuli seem to be leading candidates. Clearly, this temporal parameter is very different from the integration time-constant of 200 msec.

We have now reviewed the polar opposites of the temporal scale. Next we will analyze some other experiments concerning temporal discrimination and attempt to determine whether they are more closely related to the short or long temporal value.

THE DETECTION OF AUDITORY GAPS

That a sensory system has some inertia and that it takes some finite amount of time to respond to any change in the stimulus conditions is hardly a controversial idea. How to estimate this temporal parameter remains more troublesome. For a good review of the earlier approaches to this problem and some of their short-comings, Plomp (1964) is to be recommended. It was he who first advocated the study of the detection of silent intervals and explained how they could be used to estimate the decay or relaxa-tion of some sensory process. Figure 2, taken from Plomp, shows the essence of this approach. The decay of the sensations created by the first sound can be estimated to within a small amount, delta s, by simply turning on a second sound and adjusting its level until the silent gap, delta t, is just audible. The idea has a simplicity that recommends it, and no one has seriously challenged the concep-tion in nearly two decades. Penner (1977) pointed out, theoretical-ly, that the duration of the second sound should have little effect on the estimate of the level of decay. She showed, experimentally, that this was indeed the case. Her data almost exactly replicated Plomp's for second-sound durations of 2 and 200 msecs.

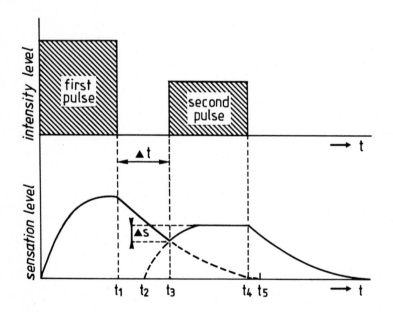

Fig. 2. Plomp's hypothesis concerning the decay of sensation and how it might be measured. The sensation caused by the first pulse decays over time. For the observer to detect a gap between the two pulses, labeled delta t, there must be a small perturbation in sensation level labeled delta s. Thus, the value of the decaying sensation can be measured to within a small amount, delta s. (Plomp 1964).

What was remarkable about Plomp's results is that they showed the
decay of sensation to be straight lines on log-log coordinates
(sensation level of the second sound versus log silent interval),
with all of the functions decaying to zero sensation level at about
200 to 300 msec, independent of the level of the first sound.
Figure 3 shows some typical results. The form of this relationship
is disturbing, since the simplest model of a decay process is an exponential one. In that case, the decay in decibels should be linear
with time (not logarithmic time). Duifhuis (1973) was one of the
first to suggest that a two-stage exponential decay could be used
to provide a reasonable fit to Plomp's data.

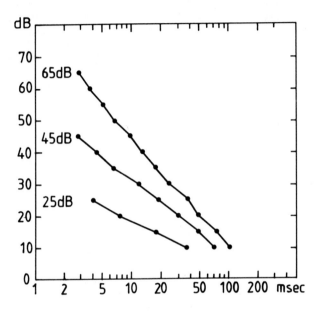

Fig. 3. The just noticeable gap between two noise bursts, see Fig.
2, as a function of the sensation level of the second noise burst.
The sensation of the first noise burst is indicated. (Plomp 1964).

When the second sound is the same level as the first, the detection
tasks amount to simply detecting a silent interval or gap between
two sounds. This simple experiment has attracted considerable attention lately because its experimental simplicity makes it useful
in a clinical assesment of hearing. Boothroyd (1973), Irwin,
Hinchcliff and Kemp (1981), Fitzgibbons and Wightman (1982), and
Irwin and Purdy (1982) were among the first to explore its use in
this connection. Several recent studies have explored the stimulus
domain in more detail. Buus and Florentine (this volume) present
more detailed information on this topic and especially its application to hearing impairment. The important physical variables can be
summarized fairly simply.

1)   Gap detection improves as stimulus level increases, at least

over the first 20 to 30 dB of sensation level and is nearly
constant at high levels (Plomp 1964, Fitzgibbons 1983, Florentine and Buus 1983, Shailer and Moore 1983).

2) Wideband signals produce the smallest gap thresholds, about 2
to 3 msec for moderately intense noises (Plomp 1964, Florentine and Buus 1983, Fitzgibbons 1983, Penner 1975).

3) Using notched noise to restrict listening to definite regions
of the spectrum, gap threshold decreases with increasing frequency up to about 1 to 4 kHz (depending on the study) and are
largely constant after this frequency (Fitzgibbons 1983,
Fitzgibbons 1984, Fitzgibbons and Wightman 1982, Florentine
and Buus 1983, Shailer and Moore 1983).

4) Detectability of the gap centered in a noise burst improves as
the duration of the noise burst is increased to about 20 msec
and remains constant for longer durations (Penner 1975).

5) The sound level in the gaps is relatively unimportant for the
measured gap threshold as long as the level is 5 to 10 dB
below the initial sound level (Irwin and Purdy 1982, Penner
1975).

All of these facts are roughly consistent with a simple energy
detection model having a first stage of filtering followed by a
square-law detector and an integrator (Florentine and Buus 1983).
The bandwidth of the first stage of filtering limits the temporal
resolution at low frequencies. At higher frequencies where the
critical band is much wider, the first stage of filtering cannot
account for the poor temporal resolution and a post-detection integrator must be blamed for the 2- to 3- msec thresholds. Buus and
Florentine (this volume) explain the application of this model and
suggest how it can be useful in interpreting data from listeners
with hearing impairment.

The close connection between the discrimination of a silent interval and the phenomenon of forward and backward masking has not gone
unnoticed (Penner 1975, Plomp 1964). Why gap detection should be
afforded this considerable interest, other than its connection with
masking, remains unclear. Can we learn something about the auditory
mechanism by discriminating a sudden termination of the sound that
could not be learned by discriminating a brief increment in intensity? What is important or interesting to be learned from a decay
in the sound that could not be learned from an increase in its
intensity? Consider a broad-band noise stimulus. In a gap detection
experiment, we ask the observer to detect a brief decrease in the
intensity of the noise. We could equally well measure the detection
of a brief increase. Would a brief increase of the same total power
be as detectable as a brief decrement of the same amount? In short,
would the minimum audible gap be equal to the minimum audible
increment? Irwin and Purdy (1982) provide us with some limited
answers to these questions. They measured the smallest times needed
to hear increments or decrements in a wide-band noise burst. For
equal increments or decrements in power, the increments are always
more detectable than decrements, especially when the changes in
level are small. For example, a 1 dB change in intensity requires a
burst duration of 50 msec to be audible, while a gap of 90 msec is
needed to hear the same decrement. For larger changes in intensity,
the difference in the temporal thresholds for increments and decrements is much smaller, although the ratio of the two times in-

creases, for example, with a 10 dB change, the threshold for an increase is 1.5 msec and for a decrease is 6 msec. However, estimates of the time constants for increments and decrements are barely different, 66 and 71 msecs. Unfortunately, we only have estimates of the asymmetry between increments and decrements in terms of temporal thresholds. No values are presently available for the threshold change in intensity for an increment or decrement of a <u>fixed</u> temporal gap. Irwin and Purdy's data suggest that this difference would be very large for short duration gaps and negligible for gaps of 10 msec or more.

When we think of measuring the detectability of increments in a sound, the natural experimental paradigm is the time-intensity trade. We know that an increase in the duration of noise increment will increase its detectability, measured in S/N ratio, up to a duration of about 200 msec (Green 1960). From this data, we conclude that the integration time constant is about 200 msec, a value often used in this area (Plomp 1964, Zwislocki 1960). There are no studies of the time-intensity trade for a decrement in noise intensity of which I am aware. Frankly, I would be surprised if the data for a decrement were greatly different from those obtained with an increment. It seems likely that the measured time-constant would be similar.

Now let us examine the exact estimates of the temporal parameters measured in these two different time experiments. From the time-intensity trade I would guess that we measure a time-constant of about 200 msec. For a loud, wide-band noise the minimal audible gap is 3 msec. Are these two numbers manifestations of the same auditory mechanism? Irwin and Purdy (1982) use Zwislocki's temporal integration theory, but find, as we have just reported, a time-constant nearly a factor of three smaller than the 200 msec value. They do not comment on this discrepancy in their article other than to say that such discrepant estimates in time-constants are often seen in this literature.

Let us consider the gap experiment from the viewpoint of an optimum detection model of a change in the level of a noise process. As can be found in Green and Swets (1966), the optimum model is an energy detector. Its detection index, d', is given by the following formula.

$$d' = (WT)^{1/2} \cdot (S/N) \qquad \text{Eq. 1}$$

Where W is the bandwidth of the noise or signal and T is the duration of the change, and S/N is the ratio of the change in noise power (in our case a decrement) compared with the average (initial) value. This equation is a good approximation as long as the degrees of freedom is reasonably large, $WT \gg 1$. For a simple integrator, the noise power decays exponentially from its initial value, N, during the gap and reaches a value, $y(T)$, at the end of the gap.

$$y(T) = N \cdot \exp(-kT) \qquad \text{Eq. 2}$$

where 1/k is the time-constant of the integrator. At the termination of the signal the average noise power returns exponentially to its initial value N. The signal is the change in noise power given by

$$S = N - N \cdot \exp(-kT) \qquad \text{Eq. 3}$$

The fluctuation on the power of the noise limits the ability to detect small changes in power. These power fluctuations are summarized by the degrees of freedom, 2WT, of the noise process. In our case, this is twice the product of the detectors bandwidth W and the time-constant of the integrator $1/k$. Thus, rewriting Eq. 1 for our case,

$$d' = (\frac{N-N \cdot e^{-kT}}{N}) \cdot (W \cdot \frac{1}{k})^{\frac{1}{2}}$$

Eq. 4

$$d' = (1-e^{-kT}) \cdot (W \cdot \frac{1}{k})^{\frac{1}{2}}$$

For detecting increments in noise power, human observers are about 5 dB poorer than optimum statistical detectors. Thus we may assume that for the human observer the observer d', d'(human) is approximately

$$d'(human) = c(1-e^{-kT}) \cdot (W \frac{1}{k})^{\frac{1}{2}}$$

Eq. 5

where c, our inefficiency parameter is about 1/3 (5 dB).

If kT is small we can approximate the exponential term by 1-kT and then we have

$$d'(human) = \frac{1}{3} (kT) \cdot (W \frac{1}{k})^{\frac{1}{2}} = \frac{1}{3} (kW)^{\frac{1}{2}} T$$

Eq. 6

for $d'=1$ we then find the time-constant $1/k$ is

$$1/k = \frac{1}{9}T^2 W$$

Since the gap threshold T is approximately $3 \cdot 10^{-3}$ seconds, the time-constant, $1/k$, is approximately

$$1/k = 10^{-6} W$$

so if we assume that W is 10.000 Hz then we have a time-constant of about 10 milliseconds. To achieve a time-constant of 100 milliseconds we are forced to make the bandwidth estimate an unrealistic 100.000 Hz. However, these calculations are sensitive to the exact value of c, the inefficiency parameter. Changing the value of c by a factor of two changes the time-constant estimate by a factor of four.

Nonetheless, we believe that gap detection is probably limited by a time constant in the shorter time range, probably 5 to 20 msec. If so, it would be close to the critical duration estimate of Penner, Robinson and Green (1972). They measure the detectability of a brief increment (a click) centered in a burst of noise. As the duration of the noise burst is varied from short to long duration, the detectability of the click first increases and then asymptotes and is independent of the noise burst duration. The duration at which the break occurs has a value of about 14 msec for wide-band (5200 Hz) noise.

This analysis strongly suggests that the gap threshold is closely related to the shorter temporal acuity value rather than the longer time-constant measured in the time-intensity trade. More exact estimates of the time constant required for gap detection will depend on more accurate estimates of the bandwidth parameter, W. It would be interesting to measure both time-intensity trade, temporal acuity, and gap thresholds for individual observers to determine the correlation between the three values. This analysis would suggest that we might find a high correlation between the gap threshold and the temporal acuity values, but little correlation with the critical duration estimate.

TEMPORAL MODULATION TRANSFER FUNCTIONS

While gap detection provides one means of studying the temporal characteristics of the auditory process, another, more general, approach is provided by linear system analysis. As the preceding discussion demonstrated, the important temporal parameter was the time-constant of an exponential decay process. Rather than look at this process in the time domain, one may also regard it as a simple low-pass filter in the frequency domain. In this case, a complete characterization of this auditory process would involve describing this filter's response as a function of frequency, that is, the filter's attenuation as a function of frequency. The advantage of this approach is that 1) to the extent the system is linear, and 2) to the extent that detection is limited by this filter, it could be used to predict the detection of any arbitrary change in the noise as a function of time by simply using the standard techniques of linear analysis.

Rodenburg (1972) and Viemeister (see Green 1973) were the first to pursue this approach. Using noise that was amplitude modulated by a sinusoid, they studied the detectability of such modulation as a function of the frequency of the modulation. The just detectable amplitude of modulation as a function of the frequency of modula- tion showed the usual low-pass characteristic, with a 3 dB down point of about 50 Hz, that is, a time-constant of about 3 msec (Rodenburg 1977, Viemeister 1977). Viemeister has also placed a click at various phase values within the sinusoidally modulated noise. The threshold for the click varies sinusoidally, as one would expect, and the amplitude of the variation diminishes with frequency as one would expect for a low-pass filter. Rodenburg (1972) and Viemeister (1979) have shown that the form of the low-pass filter is largely independent of the level of the noise. They have also shown that the breakpoint of the low-pass filter changes as a function of the center frequency of the noise band that contains the temporal modulation. Viemeister, for example, shows the time-constant changes from 5 msec at 500 Hz to 1.8 msec at 4000 Hz. This is in the same direction, but smaller in relative amount, than the change in gap thresholds as a function of fre- quency, see Fitzgibbons (1983), Florentine and Buus (1983).

Although there are many positive aspects of this linear system approach, its application to a system as complicated and non-linear as the auditory detection system requires great delicacy. As both Rodenburg and Viemeister observe in their papers, the measured thresholds often reflect much more than the attenuation values of a simple filter. To make full use of this approach, one must have considerable prior understanding of the detection process. The

interested student of this area should particularly study Viemeister's (1979) article, since it presents an extended discussion
of the factors, other than the attenuation value of the filter,
that influence the obtained data.

TEMPORAL PROPERTIES OF PITCH

Comparatively little research has been devoted to temporal properties of pitch. Henning (1970) studied how the ability to discriminate between two sinusoids of different frequency depended on the
duration for which they were presented. Frequency discrimination
improves with increases in signal duration until a duration of
about 200 msec is reached. Further increases in duration have
little effect.

A faint pitch sensation can be heard when listening to a noise
having a regular ripple in the power spectrum (Bilsen 1968). A
simple way to produce such a stimulus is to delay the noise, attenuate it, and add it back to itself. The size of the ripple is
related to the amount of attenuation in the delay path. The larger
the size of the ripple, the more salient is the perceived pitch.
Yost and Hill (1978) have suggested that the attenuation value at
which the rippled spectrum can be discriminated from a flat spectrum is a measure of "pitch strength". The duration of the noise
needed to hear this ripple in the spectrum was studied by Buunen
(1980) and the time required to discriminate one frequency of
ripple from another was studied by Yost (1980). Both found performance increased with duration up to a value of 300 to 500 msec.
Apparently, pitch strength improves as stimulus duration is increased to this relatively long time.

van Zanten and Senten (1983) have studied the perception of this
pitch when the ripple (and hence the pitch) is varied over time.
They generated a special stimulus in which the ripple (200 Hz
beteen peaks) moved with time. They studied the ability of observers to discriminate a flat noise spectrum from one in which the
ripple either 1) moved up, 2) moved down, or 3) alternated in phase
over time. When the ripple is large, moving the ripple makes the
pitch more obvious and salient. The discrimination data, however,
failed to show any superiority to these temporally varying spectra.
The difference between flat and rippled spectra was most obvious
when the ripple did not change over time. The threshold value for
the ripple was about equal for the three time-varying stimuli. The
temporal variation showed a low-pass filter characteristic. Very
fast variation was most difficult to hear and the corner frequency
was about 4 Hz. This value is about an order of magnitude lower in
frequency than the value for detecting amplitude modulation in
noise as discussed in the previous section. Thus, at least for
these stimuli, pitch perception appears to be a relatively slow
process and fixing the spectra in time promotes the best discrimination of the rippled spectrum versus a flat spectrum.

Finally, we conclude this discussion of temporal processes with a
discussion of a binaural time parameter.

BINAURAL TIME-CONSTANT

We know that the binaural system can make exquisite comparison of
the time between events occurring at the two ears. In man, for
example, one can reliably discriminate whether a click stimulus
arrives about 30 microseconds earlier at one or the other ear. If
several clicks are presented within a brief interval of time, that
time can be made about a factor of 3 smaller (Klumpp and Eady
1956). Such fine discriminations are presumably mediated by peripheral comparisons occurring at the very early stages of auditory
processing. A detailed discussion of such limits could occupy another chapter. Blauert's book (1974, 1983) provides a detailed
summary of this kind of data. Here we will briefly review a
temporal property of a more central mechanism. As a result of these
and other peripheral comparisons, the information supplied by the
two ears is interpreted by some more central process which results
in the sound source being localized in space, or, if listening with
headphones, forming what is called a lateral image. As the binaural
parameters change, the apparent image of the source moves and
occupies different conditions. How long does it take this central
process to form these binaural images and how quickly can they be
changed or altered? This is the temporal process we will consider
in this final section of the paper.

Harris and Sergeant (1971) and Perrott and Musicant (1977) found
that the minimal audible displacement of a sound source is notice-
ably poorer when the source is in motion than when two static positions are compared. Stimulated by these earlier observations, Grantham and Wightman (1979) have provided detailed information about a
central binaural time-constant. They used the fact that an antiphasic signal (Spi) is more detectable when the noise is in-phase
at the two ears (No or r=+1.0) than when the noise is out-of-phase
(Npi or r=-1.0). They sinusoidally modulated the correlation of the
noise between these two extremes and measured the detectability of
the signal at three different frequencies as a function of the rate
of modulation. If the binaural processor can follow the changing
interaural correlation of the noise, then the detectability of the
antiphasic signal would be good. If the system is sluggish and
cannot follow these dynamic changes, then detection of the signal
will suffer.

Since they could measure the quality of detection for each of several interaural noise correlations in the static conditions, each
dynamic condition could be represented by an equivalent static
correlation. The time-constant of the exponential averager was then
varied to obtain an average correlation equal to the "equivalent"
correlation obtained in the static case. In this way, binaural
time-constant estimates could be obtained at the three frequency
regions tested, 250, 500 and 1000 Hz. The estimated time constants
range from 43 to 243 msec, with the 500 Hz data showing quicker
response than the 250 Hz condition. Comparison with the 1000 Hz
condition was less certain, in large part because of the size of
the effect, and hence the potential error of measurement was much
greater.

This result is of special interest because these estimates of the binaural time-constant are very close to that obtained with more traditional critical durations, as estimated from time-intensity trades. This result is somewhat unexpected because Grantham and Wightman's task is essentially an acuity task. As with detection of a temporal gap in noise, the observer will do better the smaller the value of the time-constant. This result suggests, at least for this system, that a single time-constant in the 200 msec range may suffice to describe all important aspects of binaural detection and localization process.

SUMMARY

As indicated in the introduction, there are a variety of different temporal parameters that one can find in modern psychoacoustic investigations. For intensity discrimination tasks, there is general concensus on a long time-constant, about 200 msec, found in studies measuring the detection of weak signals in either noise or quiet. There also appears to be a minimal time-constant of about 2 msec that is revealed in temporal acuity tasks, although there is some difference of opinion on the very shortest times that one can reliably discriminate differences in the temporal order of a pairs of clicks. I have argued that gap detection is more closely related to the acuity value, but exact estimates depend on better estimates of the bandwidth parameter, W. It may be that gap detection reveals still another temporal parameter, somewhat larger than the acuity value. It would appear that temporal parameters measured with temporal modulation techniques are also in the shorter range, about 2 to 3 msec. A few studies have been carried out on auditory abilities not directly related to intensity discrimination or temporal acuity. Those in the area of pitch suggest the pitch processor is fairly slow, and the binaural mechanism responsible for locating the binaural image is in the 200 to 500 msec region. While it would be exciting to know that one parameter represents peripheral processing and another parameter represents a central process, I believe such inferences are premature. There is still enough uncertainty about estimates made in a single task to caution such speculations. I would think a more sensible task is to try to understand the relationship, if any, between the various temporal estimates we presently have acquired in different psychoacoustic tasks.

ACKNOWLEDGEMENTS

The author gratefully acknowledges the support of the National Science Foundation and the National Institutes of Health. I also wish to thank Dr. Thomas Hanna and Christine Mason who read and commented on parts of this article. I also would like to thank Ms. Dorothy Riehm who provided technical assistance in the preparation of the manuscript. Dr. Søren Buus is especially to be thanked since he found an error in my original derivation of the optimum noise detector.

REFERENCES

Bilsen FA (1968) On the interaction of a sound with its repetitions. Thesis, Delft University of Technology

Blauert J (1974, 1983) Spatial Hearing: The Psychophysics of Human Sound Localization, trans JS Allen. MIT, Cambridge, Massachusetts (original edition published by SH Verlag, Stuttgart 1974)

Boothroyd A (1973) Detection of temporal gaps by deaf and hearing subjects (SARP No 12). Clarke School for the Deaf, Northampton

Buunen TJF (1980) The effect of stimulus duration on the prominence of pitch. In 5th International Symposium on Hearing, ed G van den Brink, Delft UP, Delft, pp 374-378

Duifhuis H (1973) Consequences of peripheral frequency selectivity for nonsimultaneous masking. J Acoust Soc Am 54:1471-1488

Feth LF, O'Malley H (1977) Influence of temporal masking on click-pair discriminability. Percpt Psychophys 22:497-505

Fitzgibbons PJ (1983) Temporal gap detection in noise as a function of frequency, bandwidth, and level. J Acoust Soc Am 74:67-72

Fitzgibbons PJ (1984) Temporal gap resolution in narrow-band noises with center frequencies from 6000-14000 Hz. J Acoust Soc Am 75:566-569

Fitzgibbons PJ, Wightman FL (1982) Gap detection in normal and hearing-impaired listeners. J Acoust Soc Am 72:761-765

Florentine M, Buus S (1983) Temporal acuity as a function of level and frequency. In Proceedings of the International Congress of Acoustics, 11th Meeting, Paris 1983, pp 103-106

Grantham DW, Wightman FL (1979) Detectability of a pulsed tone in the presence of a masker with time-varying interaural correlation. J Acoust Soc Am 65:1509-1517

Green DM (1960) Auditory detection of a noise signal. J Acoust Soc Am 32:121-131

Green DM (1971) Temporal auditory acuity. Psychological Review 78:540-551

Green DM (1973) Minimum integration time. In Basic Mechanisms in Hearing, ed AR Møller. Academic Press, New York, pp 829-846

Green DM, Swets JA (1966) Signal detection theory and psychophysics. Wiley, New York (Reprinted RE Krieger, Huntington NY 1974)

Harris JD, Sergeant RL (1971) Monaural/binaural minimum audible angles for a moving sound source. J Speech Hear Res 14:618-629

Henning GB (1970) A comparison of the effects of signal duration on frequency and amplitude discrimination. In Frequency Analysis and Periodicity Detection in Hearing, ed R Plomp and GF Smoorenburg. AW Sijthoff, Leiden, pp 350-361

Henning GB, Gaskell H (1981) Monaural phase sensitivity with Ronken's paradigm. J Acoust Soc Am 70:1669-1673

Huffman DA (1962) The generation of impulse-equivalent pulse trains IRE Transactions, IT8, pp 510-516

Irwin RJ, Hinchcliff L, Kemp S (1981) Temporal acuity in normal and hearing-impaired listeners. Audiology 20:234-243

Irwin RJ, Purdy SC (1982) The minimum detectable duration of auditory signals for normal and hearing-impaired listeners. J Acoust Soc Am 71:967-974

Klumpp RG, Eady HR (1956) Some measurements of interaural time difference thresholds. J Acoust Soc Am 28:859-860

Leshowitz B (1971) The measurement of the two-click threshold. J Acoust Soc Am 49:462-466

Munson WA (1947) The growth of auditory sensation. J Acoust Soc Am 19:584-591

Patterson JH, Green DM (1970) Discrimination of transient signals having identical energy spectra. J Acoust Soc Am 48:894-905

Penner MJ (1975) Persistence and integration: Two consequences of a sliding integrator. Percept Psychophys 18:114-120

Penner MJ (1977) Detection of temporal gaps in noise as a measure of the decay of auditory sensation. J Acoust Soc Am 61:552-557

Penner MJ, Robinson CE, Green DM (1972) The critical masking interval. J Acoust Soc Am 52:1661-1668

Perrott DR, Musicant AD (1977) Minimum auditory movement angle: binaural localization of moving sound sources. J Acoust Soc Am 62:1463-1466

Plomp R (1964) Rate of decay of auditory sensation. J Acoust Soc Am 36:277-282

Plomp R, Bouman MA (1959) Relation between hearing threshold and duration for tone pulses. J Acoust Soc Am 31:749-758

Resnick SB, Feth LL (1975) Discriminability of time-reversed click pairs: intensity effects. J Acoust Soc Am 57:1493-1499

Rodenburg M (1972) Sensitivity of the auditory system to differences in intensity. Doctoral thesis. Medical Faculty of Rotterdam

Rodenburg M (1977) Investigation of temporal effects with amplitude modulated signals. In Psychophysics and Physiology of Hearing, ed EF Evans and JP Wilson, Academic Press, London, pp 429-437

Ronken DA (1970) Monaural detection of a phase difference between clicks. J Acoust Soc Am 47:1091-1099

Scharf B (1978) Chapter 6, Loudness. In Handbook of Perception Vol. IV, ed. Carterette and Friedman. Academic Press

Shailer MJ, Moore CJ (1983) Gap detection as a function of frequency, bandwidth, and level. J Acoust Soc Am 74:467-473

Small AM, Brandt JF, Cox PG (1962) Loudness as a function of signal duration. J Acoust Soc Am 34:513-514

van Zanten GA, Senten CJJ (1983) Spectro-temporal modulation transfer function (STMTF) for various types of temporal modulation and a peak distance of 200 Hz. J Acoust Soc Am 74:52-62

Viemeister NF (1977) Temporal factors in audition: A system analysis approach. In Psychophysics and Physiology of Hearing, ed EF Evans and JP Wilson, Academic Press, London, pp 419-427

Viemeister NF (1979) Temporal modulation transfer functions based upon modulation thresholds. J Acoust Soc Am 66:1364-1380

Watson CS, Gengel RW (1969) Signal Duration and Signal Frequency in Relation to Auditory Sensitivity. J Acoust Soc Am 46:989-997

Yost WA (1980) Temporal properties of the pitch and pitch strength of ripple noise. In 5th International Symposium on Hearing 1980, ed G van den Brink. Delft UP, Delft, pp 367-373

Yost WA, Hill R (1978) Strength of the pitches associated with ripple noise. J Acoust Soc Am 64:485-492

Zwislocki JJ (1960) Theory of temporal auditory summation. J Acoust Soc Am 32:1046-1060

Zwislocki JJ (1969) Temporal summation of loudness: An analysis. J Acoust Soc Am 46:431-441

Auditory Time Constants: A Paradox?

E. de Boer

Academic Medical Centre, Meibergdreef 9
1105 AZ Amsterdam, the Netherlands

ABSTRACT

Three types of experiments on temporal discrimination are considered. The first is threshold detection of tone bursts with the
duration of the burst as the independent variable. Results can be
described by a model incorporating a "leaky integrator" which
operates on the intensity of the signal. The appropriate time constant is 200 msec. Next comes the detection of amplitude modulation. The threshold found as a function of modulation frequency
(for low modulation frequencies) can again be described by a "leaky
integrator" model, this time equipped with a time constant of 20
msec. It is argued why these two models are incompatible: the auditory system appears to be able to detect modulations far easier
than the 200 msec time constant model would allow.

More directly aimed at temporal phenomena are experiments on forward masking and the detection of gaps in a continuous signal.
Interpretation of the findings is rather difficult because there is
a peculiar nonlinearity in the temporal persistence of masking and
because the "ringing" of the auditory filters is difficult to
separate from the persistence of masking. It is argued why, on the
basis of experimental results, one should come to the conclusion
that the auditory system is capable of detecting a 2-4 dB decrement
of intensity in approx. 10 ms. This is compatible with what we can
infer about the ringing of the filters. However, it is incompatible
with the 200 msec time constant model of temporal integration: the
auditory system is far less capable of detecting increments than
decrements. The conclusion is that each type of experiment leads to
a model describing only the results of that experiment. The three
models described should be regarded as "ad hoc" models because they
cannot be united into one model.

INTRODUCTION

Frequency and time are traditionally the parameters to vary in
order to study the primary properties of our hearing organs. In the
classical work in psychophysics the concept of the critical band
was associated with the frequency variable and the time constant of
auditory integration with the time variable. The processes of filtering and integration could be considered as nearly independent.
In modern work, where stimuli are used that are closer to the

sounds involved in typical auditory tasks, fairly complex interactions between frequency and time have come to light. So complex, in fact, that is has become difficult to understand the operation of the auditory system as a whole.

Models are the instruments to facilitate understanding. They do this by explaining the relation between experimental results in a rational way. It would be fruitful if such models would definitely correlate with physiological mechanisms but this is not always the case (and is not always possible to achieve). As a matter of fact, models are often conceived so as to encompass only a limited set of experiments or a single phenomenon. The filter-bank model, for instance, deals mainly with spectral resolution and not with temporal variations. Similarly, the integrator model only describes auditory thresholds for stimuli of different durations and cannot treat the dynamics of sounds. One model cannot explain phenomena that lie in the domain of the other, and vice versa. This is a deplorable situation, the more so since "meaningful" sounds necessarily include variations in frequency and time.

In auditory theory there is, at present, not one model - a super-model, so to speak - from which the aforementioned models (and all related models) constitute a subset. Moreover, existing models turn out to be partly contradictory if one would attempt to integrate them into a super-model. It is the purpose of this paper to point out instances in which this deficit becomes clear. Or, in other words, situations are considered for which existing models are contradictory or not rational if viewed in a wider framework.

To clarify the point at issue it is neccessary to briefly review aspects of linear filtering and temporal integration first. This is done in sections "Peripheral filtering, spectral and temporal effects" and "The auditory integrator", respectively. On this basis some fundamental results of classical and recent studies on temporal phenomena are reviewed, in sections "Modulation detection", "Temporal tracings, forward masking" and "Detection of temporal gaps". The final section describes the basic flaws in the treated models and calls for ideas to be used in attempts to integrate existing models in a meaningful way.

PERIPHERAL FILTERING, SPECTRAL AND TEMPORAL EFFECTS

In the physics of sound, the concept of frequency is used in two ways:
1) as expressing the rate of periodic repetition and
2) as the independent variable of the Fourier spectrum.
In mathematical terms the distinction is clear but as far as the relevance for auditory perception (psychophysics) is concerned, this is not so clear. For example, it has taken more than 130 years to come to a more or less coherent view as to which type of frequency is responsible for which type of pitch (for a review see de Boer 1976). Yet the general role of frequency in the spectral sense is fairly well established. In many respects the resolution in (spectral) frequency is governed by the critical band and this,

in turn, is considered to be due to peripheral filtering of the auditory stimulus. Many (but certainly not all) auditory phenomena have properties suggesting that the filtering process is nearly linear and we will be concentrating on linear filtering in what follows.

The critical-bandwidth concept went through several stages since its conception (Fletcher 1940). It was used at first to describe auditory integration over frequency, later on it was used also to describe spectral resolution. That is, interest shifted from the width of the critical band to the slopes of the critical-band filter (Patterson 1976). Still later, temporal aspects became important too - these will be treated further on. Psychophysical measurement of the critical band and, particularly, of the associated filter response curve is very difficult and tricky because there are many confounding factors. In the course of time the following have been recognized: nonlinearity, the influence of fluctuations inherent in noise stimuli (de Boer 1966; Bos and de Boer 1966; Patterson and Henning 1977), off-frequency listening (O'Loughlin and Moore 1981) and frequency splatter, temporal factors (the "overshoot" effect, see Zwicker 1965), two-tone suppression (Moore and Glasberg 1982). Much thought has been given to the question whether the critical-band mechanism requires time to be built up or not. Current thinking favours the latter possibility, it is as if the filtering mechanism is instantaneously available as soon as a signal is switched on (Zwicker and Fastl 1972). Many physiological data, notably those on tuning curves, suggest that critical-band filtering is established mechanically in the cochlea although the actual mechanics is incompletely understood (Khanna and Leonard 1983).

As indicated earlier, peripheral auditory filtering has many properties akin to those of linear filters and, if this is true, the associated temporal properties should make themselves felt. The waveform of a filtered signal cannot vary arbitrarily fast in amplitude. Let us, to put this into more concrete terms, consider first the simplest possible filter, a single resonator. This can assume the form of an electrical circuit with inductance L, capacitance C and resistance R as in fig. 1. Input is the voltage E,

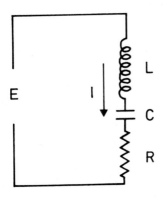

Fig. 1. Simple resonator

sinusoidal with radian frequency ω and amplitude a, output is the current I through the circuit.

For fixed a, the current I has maximum amplitude when the frequency $\omega/2\pi$ is equal to the <u>resonance frequency</u> $\omega_o/2\pi$ of the circuit where ω_o has the property $\omega_o^2 LC = 1$. For simplicity we assume the circuit to be dimensioned in such a way that ω_o equals 1. Resonance exists not only at exactly the radian frequency $\omega_o = 1$ but extends over a certain range on both sides of the resonance frequency. The <u>bandwidth</u> β of the circuit is commonly taken between the points where the amplitude of I is 3 dB below the maximum amplitude. For the circuit shown, and the parameters L and C chosen such that LC = 1, the bandwidth (in radians per second) is given by β = R/L.

When E is suddenly made zero, the current I continues to oscillate for a while - at the resonance frequency - but its amplitude decays exponentially with time toward zero according to the function exp $(-t/\tau_o)$. The value of the <u>time constant</u> τ_o is 2L/R. Hence there is a simple relation between the time constant τ_o and the bandwidth β:

$$\tau_o = 2/\beta \qquad (1)$$

This remains true when frequencies are scaled to the range characteristic for the ear: when B is the bandwidth in Hz, the associated time constant τ_o (in seconds) is:

$$\tau_o = 1/\pi B \qquad (2)$$

This holds true for a single resonance. It is not true for a more complex resonator. Auditory filters have relatively flat response curves near their peaks but extremely steep skirts. This is certainly related to the analogous property exhibited by auditory-nerve fibre responses. Tuning curves (or Frequency Threshold Curves as they are often called) of auditory-nerve fibres have bandwidths on the order of 0.1 - 0.4 times the characteristic frequency but skirts with extremely high slopes, 100 - 1000 dB/octave. Such response characteristics are incompatible with a single resonator. In several respects the filters in the cochlea are equivalent to a cascade of 4 - 10 single resonators with identical properties (de Boer 1975, 1979). The temporal behaviour is then no longer exponential and it can no longer be described by a time constant. In the case just mentioned where sinusoidal excitation is suddenly switched off, the response continues to oscillate with nearly the full amplitude, and starts to decay after a certain delay. Despite this deviant behaviour, the first part of the over-all decay can approximately be described by a delay plus an exponential decay with a modified time constant. Or, still simpler, we can use a single effective time constant which describes in which time the amplitude decays to 1/e of its initial value. For a cascade of n resonators the value of this effective time constant is approximately 1.4(n-1) times the time constant associated with a single resonator. In terms of the bandwith B, the "time constant" τ_o of the peripheral auditory filters is

$$\tau_o = k/B \qquad (3)$$

where k ranges from 1.3 to 4.2.

As stated earlier, psychophysical measurement of the shape of the frequency response of the critical-band filter is tricky. It is often done via the technique of masking. Quite popular has become the technique leading to psychophysical tuning curves (Zwicker 1974): the level of the test tone is kept constant and the level (or frequency) of a second tone, masker tone, is varied to find the points where the signal is just masked. Because of the non-linearities involved, interpretation of the curves is not unequivocal although formal models can be set up to account for the data (Verschuure 1981). To isolate effects due to two-tone suppression psychophysical tuning curves are also measured with forward masking (e.g. Moore 1978), we will briefly come back to this technique later on. In this case, too, there are confounding effects that interfere with proper interpretation but this is not the place to expound on them.

In the 1970's the measurement of auditory frequency selectivity has acquired a new dimension since it became known that the width of the auditory filter is enlarged under pathological conditions. This was found in animal experiments, in temporary disfunction (Evans 1975) as well as in permanent damage (Harrison and Evans 1977), and the same type of pathology was found in humans with sensorineural hearing loss (Wightman et al. 1977). For this reason the interest in psychophysical tuning curves has increased. Unfortunately, the many factors involved in interpretation and the great variability in the data obtained from inexperienced patients have impeded general acceptance of this type of measurement.

THE AUDITORY INTEGRATOR

One aspect of the temporal behaviour of the auditory system can be described in quite simple terms: temporal integration. Short tones need a larger intensity than long tones in order to be detected. In fact, for tones shorter than 200 msec, and at their thresholds, the product of intensity and duration is constant. Or, in other terms, such tones need a certain energy (the product of intensity and duration) in order to be detected. This property of the auditory system, well-known for many years (Plomp and Bouman 1959) and well-studied from many sides (Zwislocki 1960), is commonly described in analogy with a model in which the intensity is integrated with respect to time. When a tone burst is presented, the output of the model starts to rise linearly with time. If, at or during the tone burst a certain threshold level is reached, that burst is detected. To account for the observed fact that thresholds do not improve for tone bursts longer than approx. 200 msec, the integrator is turned into a "leaky integrator" with a time constant (τ_{int}) of 200 msec. Note that the model holds true for absolute as well as masked thresholds provided spectral cues for detection are eliminated.

Popular as this model may be, it has several features that are illogical. The main question is: why is the integrator turned into action when a tone burst is presented? We can certainly discriminate events on a much smaller time scale than 200 msec - we shall see further on that a kind of universal limit occurs at 2 - 3 msec - and we know for certain that the peripheral filters are much faster. The model does not explain why the integrator would only be turned into action near threshold and why much faster detection mechanisms would be active at higher levels. To put it in straight forward terms: the integrator model is an ad hoc model, it is meant only to describe one type of measurement. It doesn't do more and thus it cannot be brought into contact with different experiments.

To illustrate this point further we should consider what goes on in the auditory periphery when long and short tones are presented. Let us take the case of a differential threshold for intensity. For long bursts of tone (or noise) the threshold corresponds to a level variation of approx. 1 dB. That means that the information carried to the brain by the fibres of the auditory nerve over a duration of 200 msec is certainly accurate enough for the central auditory system to make this discrimination. Remember that the firings of auditory-nerve fibres are essentially stochastic in character. If the statistics of the firings would be the only decisive factor, one might expect differential thresholds to vary with the square root of the burst duration. On going from 200 to 10 msec the differential threshold would then go from 1 to 6.5 dB. Actual measurements show the differential threshold to vary less, to maximally 2.5 dB (Zwicker 1956). Hence we conclude that the statistics of firings is not the main factor which limits the performance for differential thresholds. Note that this conclusion runs counter to the general ideas implied in Siebert's theory on statistical limits of auditory discrimination (Siebert 1968). In point of fact, that theory takes only into account contributions to differential thresholds due to fibres that are not saturated. We now know that saturated fibres can respond differentially to an increment of intensity (Smith and Brachman 1982). When that effect is taken into account we could expect that the differential threshold for a 10 ms burst would be less than 6.5 dB higher than that for a 200 ms burst. Yet, what the experiments on threshold detection tell us is quite a different story: a 10 ms burst has a threshold 13 dB higher than a 200 ms burst. There seems to be a wide gap between detection of a signal and detection of the difference of two signals. In terms of statistical detection theory these processes are equivalent. This short excursion should be sufficient to illustrate why we should consider the temporal integration model as an ad hoc model and why we should not try to connect it with models describing other phenomena. The excursion illustrates also the importance of being earnest about the final decision process in an auditory model and of not indulging in the wildest speculations.

MODULATION DETECTION

A convenient and not too complicated way of studying temporal aspects of hearing involves the use of modulated stimuli. A steady tone or noise signal is amplitude modulated (usually sinusoidally) and the threshold of detection of the modulation is measured as a function of the frequency of modulation f_{mod}. A confounding feature of this technique is that the modulation modifies the spectrum. One way of avoiding spectral variations to affect the results is to use a wide band of noise as the signal to be modulated. The big disadvantage of this technique, of course, is that it remains unknown which spectral region of the stimulus contributes the most to the results. We shall come back to the modulation of narrow-band signals later.

Experimental findings are usually plotted as thresholds of detection versus modulation frequency. Let us consider the case of amplitude-modulated wide-band noise (Viemeister 1977; Rodenburg 1977). The modulation threshold turns out to be smallest around $f_{mod} = 5$ Hz (Zwicker 1956). For lower modulation frequencies the threshold increases a little - this effect appears to be dependent upon the experimental technique: it is absent when a two-interval forced-choice experiment is performed (Rodenburg 1977). For modulation frequencies above 8 Hz the threshold of modulation detection increases at an asymptotic rate of 3 dB per doubling of f_{mod}. It is as if modulations with higher frequencies are attenuated by the auditory system, one often speaks of the "modulation transfer function" of the system (this function, then, would go down at the same rate as the thresholds are going up).

Again this type of behaviour can be modelled by a "leaky integrator". The modulations of sound intensity are integrated and the threshold is reached when the peak-to-valley difference exceeds a certain amount (for small f_{mod} 0.5 - 1 dB). To account for the existence of a corner frequency the integrator is endowed with a time constant τ_{mod}. This time, the time constant has to be chosen approximately equal to 20 ms (published values range from 10 to 25 ms), a factor of 10 smaller than the time constant τ_{int} for temporal integration.

It should be stressed that this leaky integrator model for modulation is again an ad hoc model. To illustrate this, we shall discuss how much the two models differ. This treatment necessarily has to be somewhat crude since we are trying to compare two dissimilar experiments. The time constant τ_{int} of 200 ms discussed in section "The auditory integrator" results from experiments with bursts of tone or noise with a (nearly) rectangular envelope. A tone with a duration of 200 ms has a threshold 1.5 dB higher than one that is considerably longer. That means that the integrator has reached about 0.7 of its ultimate output value in 200 ms. The time to reach (1 - 1/e) times the ultimate value should be taken as somewhat smaller than 200 ms, say, 180 ms. It is the latter value that should be ascribed to the integrator when it is attempted to use the same concept for sinusoidal intensity modulation. Then, for a modulation frequency of 8 Hz - the corner frequency mentioned above - the output of the integrator would be 9.6 dB down. Clearly too much in view of the data on modulation: 1.5 dB would be sufficient. Apparently the detection of modulations is some 6 dB better than that allowed for by the threshold integration model of section "The auditory integrator". In other words, resolution is better for periodic modulations than for single bursts of signal. Again we find this tentative conclusion to be in line with the property that the physiological mechanisms of the auditory periphery are capable of providing more information than is made use of by central processes. And again it remains unclear why this is so and why the system should turn to different procedures when it has to do different tasks, modulation detection or threshold detection. In particular, it is difficult to see why performance of an observer would be poorer in the latter case.

A few words are in order about frequency effects. The modulation experiment can be done with tones or narrow bands of noise as carriers. For small modulation frequencies the behaviour is the same in all cases (Zwicker 1956) and can be described by the leaky integrator model with a time constant of approx. 20 ms. There is a small effect of bandwidth for the case where a narrow-band signal is modulated (Rodenburg 1977). For sinusoidal carriers the threshold of modulation first increases with f_{mod} and then starts to decrease, this occurs at values of f_{mod} that are directly related to the value of the carrier frequency (Zwicker 1956). This effect is due to the critical-band filtering process: modulated signals can be described in spectral terms and in this case the outer sidebands are spectrally resolved. In the days when there was much uncertainty about the value of the critical bandwidth the modulation experiment was one of the methods employed to find the "true" values (Feldtkeller and Zwicker 1956, 1967).

Concommitant with these variations, one can observe a gradual variation of the roughness of the sound, at least when a larger modulation depth is employed. Starting from the smallest values of f_{mod} one observes slow variations of loudness, for larger f_{mod} the sound becomes rough but for still larger values of f_{mod} where spectral analysis sets in, roughness decreases and the sound can become quite smooth. Roughness appears to be a direct indicator of how the human ear can follow temporal variations. Roughness can be measured and expressed numerically and the experiments give a quantitative expression of the just-described phenomena (Fastl 1977).

TEMPORAL TRACINGS, FORWARD MASKING

A direct way of tracing what goes on in the time domain is the temporal probe technique (Viemeister 1973, 1977; Zwicker 1973; Green 1973; Rodenburg 1977). A probe signal is employed that is sharply defined in time, and its masked threshold is measured as a function of its time relation to the masking stimulus. Typically, the masking signal can be an amplitude-modulated wide-band noise signal and the probe a 1 ms burst of an 8.5 kHz tone (Fastl 1976). Note that the frequency of the tone is chosen so high that the appropriate auditory filters can follow the amplitude variations of the tone burst faithfully (compare section "Peripheral filtering, spectral and temporal effects"). For amplitude modulation with a large modulation index (100%) and a modulation frequency of 5 Hz

the probe threshold is observed to follow the modulations completely. For 10 Hz modulation 8 - 12 dB peak-to-valley ratio is
left and for 100 Hz only 5 dB (Fastl 1976). A closer look indicates
that the simple integrator model discussed in the preceding section
is not sufficient to explain the data, either the integrating
"filter" has to be of the 3rd of 4th order or some degree of
smearing-out has to be introduced. Note that in this case the
measurements allow the full time function to be determined so that
the associated transfer function is found with respect to its
amplitude as well as its phase. The name "modulation transfer
function" is certainly more appropriate in the present case than in
that of the preceding section where this concept is commonly used.
Remarkably enough, the phase changes remain unexpectedly small for
small f_{mod} and rise rapidly at larger values of f_{mod} (Viemeister
1977); this feature causes the aforementioned difficulties if one
wants to represent it in terms of a model.

The temporal probe technique produces more detailed results than
the simple determination of modulation thresholds as described in
the previous section. It can not only be employed when there is a
periodic variation of the intensity (as is the case with sinusoidal
amplitude modulation) but also when more complex variations are
involved. One instance is reported by Zwicker (Zwicker 1973): a
pure tone is modulated in amplitude by a pseudo-random noise signal
(a periodic signal that shows variations having most of the properties of a true random signal) and the temporal probe technique is
employed to find out to which degree the auditory system can
"follow" the amplitude variations. The result is that most details
of the modulating waveform can be followed (the probe signal in
this case is a 5 ms 4 kHz tone burst so that information about
variations occurring within 5 ms is lost).

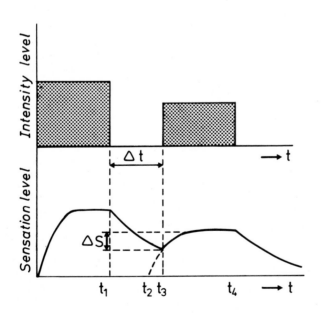

Fig. 2. Measuring the decay
of auditory sensation
(after Plomp 1964)

Another most interesting type of variation is the sudden cessation of the stimulus. Here we enter the domain of <u>forward masking</u>: the masking signal is switched off and the persistence of masking is traced by way of a probe presented after the switching-off instant. The principal point here is that the effect of the masker does not decay exponentially with time as would be the case if a single resonator were present. Hence a plot of effective level (in dB) versus time (linear) is not a straight line. By a most ingenious experiment Plomp (1964) has shown that the decay occurs nearly according to a straight line provided the time axis is plotted logarithmically (the effective level is expressed in dB as usual). Because of the extremely fast decay of masking there would be a considerable influence of the duration of the test tone burst. This problem was circumvented by Plomp in the following way. Two sound bursts are given, both of a duration sufficient to give maximum loudness and maximum masking power, 200 ms. The second burst was started Δt sec after the first one was stopped, the second burst was weaker than the first. Fig. 2, borrowed from Plomp's paper, illustrates the idea behind the experiment. Following Plomp's description: "we may assume that starting a second sound pulse at t = t_3 affects the decay curve of the first as plotted .... It is reasonable to suppose that the interruption between the pulses can be perceived only for ΔS values exceeding a critical amount. It is reasonable, too, that this just-noticeable ΔS may be identified with the difference limen of intensity" (end of citation). Just the reference to the latter point would be sufficient reason to quote this experiment here but there is also a logical bridge to the topic of the next section: gap detection. Returning to the decay of auditory sensation, that decay is not measured by finding out how much a sound probe is masked but by finding out how the sensation can be continued at a lower level without an audible gap.

The results of Plomp's experiment confirmed earlier data and have become almost "classical" in their scope and meaning: plotted logarithmically, the just-noticeable time interval is linearly related to the sensation level (SL) of the second sound burst. Interpreted in terms of an after sensation: the after sensation decays linearly with log time to zero in a constant time, approx. 200 ms. This decay time is independent of the masker level which implies that the after-effect of a strong signal decays faster to zero than that of a weak signal.

Earlier data using short signal probes had indicated the same tendencies but were not all in mutual agreement. The reason lies in the fact that masking decays so fast that the duration of the test signal becomes important. In fact, it is the last part of the test signal that is responsible for detection, the first part is masked. Accordingly, the time variable τ should not be the time from the end of the masking burst to the beginning of the test signal but from the end of the masking burst to the end of the test signal. Several later authors have followed Plomp's advice. Specifically, Fastl (Fastl 1976) showed that, provided the time interval from-end-to-end is kept constant, the amount of masking (level of the masked tone minus the level at threshold) is nearly independent of the duration of the test signal.

Techniques of forward masking have also become important in the measurement of auditory frequency selectivity, this is sufficient reason to go directly to the case where a masking tone of one frequency is followed by a short test tone of a different frequency. One important concept in this area is the masking function, i.e. the function expressing the relation between masker level and amount of masking. In the case of simultaneous masking this function has, as is well known, a slope nearly equal to 1.0. That is, in the main domain of masking, a 10 dB stronger masker produces 10 dB more masking. This rule holds when the frequencies of masker and test tones are the same. The slope is larger than 1.0 for strong maskers and test frequencies above the masker frequency, this effect can be attributed to the production of harmonics of the masker tone. There are also cases where the slope is smaller, for instance, as a result of combination tones.

In very general terms the same effects are found in forward masking but the main point is that the basic slope of the masking function is decidedly smaller than unity (Houtgast 1974). Widin and Viemeister (1979) devoted a study to the slope of the masking function with the specific aim to apply the results to the interpretation of psychophysical tuning curves determined with forward masking. They found, when the masker frequency f_m equals the signal frequency (test-probe frequency) f_s, that the slope is smaller than 1.0 and depends considerably on the time interval τ between masker and signal tones. When $f_m < f_s$ the slopes are somewhat larger (as expected) and for $f_m > f_s$ they are sometimes smaller. There are, in this respect, rather large inter-individual differences and listeners may go through a fairly long learning process. Because of these two effects caution is necessary in the interpretation of psychophysical tuning curves measured in forward masking. In any event, it is stressed to consider each tuning curve individually and not to average across subjects lest important details should get lost.

Psychophysical tuning curves measured with forward masking are generally sharper than those obtained with simultaneous masking. On the high-frequency side of the masker frequency slopes of 310 - 650 dB/octave are reported by Moore (1978), 560 dB/octave by Vogten (1978) and 240 - 370 dB/octave by Widin and Viemeister (1979). Low-frequency slopes range from 45 to 120 dB/octave. All these values were obtained with forward masking. Slopes in simultaneous masking are invariably smaller, the difference is usually attributed to suppression phenomena (Moore and Glasberg 1981). Since the suppression effect is entirely outside the scope of the present paper we will not go into any detail here. When off-frequency listening is made impossible, the slope values are reduced considerably. At 1 kHz the shape of the critical-band filter's response curve for simultaneous masking is equivalent to the value 2 for n (the number of simple resonators). For forward masking a value of 4 would be more appropriate. This tallies with the value found from tuning curves of auditory-nerve fibres (see section "Peripheral filtering, spectral and temporal effects").

The fact that in forward masking the slope of the masking function is less than unity is interesting in itself. It betrays a nonlinear effect that is quite distinct from the nonlinearity involved in auditory filtering and in the production of harmonics. It is not a simple matter to model this type of nonlinearity since it does not seem to fit with one of the known theories of auditory discrimination. The same applies to the decay of masking as a function of time, these two properties are tightly interwoven, of course, one cannot exist without the other.

A parametric study of forward masking has been carried out by Jesteadt and co-workers (Jesteadt et al. 1982). The argument was that at least two types of time constants should be necessary to describe effects of masker frequency, masker intensity and time interval τ. The results turned out simpler than expected, only three parameters are sufficient to describe the trends of all results. In particular, no dichotomy in the dependence on was noted despite the fact that a masker frequency as low as 125 Hz was included. Mainly on the basis of analogies with physiological data the authors argue that the principal influence of frequency goes via the slope of the masking function and that the properties of the recovery process are basically the same for all frequencies (i.e. for all locations in the cochlea).

For the purpose of the present paper it would be desirable to make a connection between masking during the last few ms of the masker burst and that in the first few ms of the silent interval after the masker burst. Unfortunately, the data are given in such a way that this is impossible. The use of a logarithmic time scale precludes plotting of data at values of τ that are zero or negative. Moreover, not all authors define τ from end-to-end as they should (see what has been said above about Plomp's work). Only Widin and Viemeister report that for the smallest τ-values their data asymptote at a value below the simultaneous threshold. The scarcity of data makes it impossible for us to infer what is actually happening near the end of the masking burst; in particular, it is not clear on what type of variation the detection of the signal probe is based. Thus it is, once more, difficult to build a bridge between different types of experiment and between different theories or models.

DETECTION OF TEMPORAL GAPS

Let us first recall the temporal-probe technique mentioned in the beginning of the previous section. A probe signal that is sharply defined in time is presented after cessation of a masking burst and the threshold of the former is measured as a function of the time interval τ. Next, add a second masking burst, of the same intensity, at a somewhat later instant. In effect we are now measuring what happens in a temporal gap of the stimulus. The tracing of signal threshold versus τ shows two effects: forward masking exerted by the end of the first masking burst and backward masking extending over a few ms before the start of the second masking burst. The main contribution to backward masking results from the peripheral filter as has convincingly been shown by Duifhuis (1973). The contribution of filtering to forward masking is generally considered as smaller than the temporal decay of masking but should certainly not be neglected.

As detailed in the previous section, when filtering effects dominate, they do not markedly affect the shape of the temporal decay. That is, there is no break in the decay function. Now we can pose the following question: if we use the smallest possible gap widths, is the persistence of masking mainly due to temporal effects of filtering or not? Let us, on the basis of the equations presented in section "Peripheral filtering, spectral and temporal effects", predict what happens if masking instantaneously follows the output of the peripheral filter. As explained in section "Peripheral filtering, spectral and temporal effects" this output does not strictly follow an exponential decay function but over the first few ms or so the trend of the decay can well be described by a modified time constant, namely, the one given by eq. (3). In a time τ_o the amplitude varies by a factor of e, i.e. by 8.7 dB. Variations of the order of 1 dB should occur in about one eigth of τ_o. Taking the situation one step further: if measurable excitation changes by an amount that is detectable by itself we would not need the signal probe any more and we have simply a case where the presence or absence of the gap can be detected. In this way we are led to the subject of <u>gap detection</u>.

Let us turn to the data on this subject. First, in doing such intricate measurements it is necessary to avoid effects of energy splatter. The introduction of a temporal gap in a continuous tone or narrow band of noise produces extraneous spectral components that can lead to detection of the gap. To avoid this problem, Fitzgibbons and Wightman (1982) added a continuous wide-band masking noise to their stimuli. This masking noise produced enough masking outside the main band of the masking noise to eliminate spectral splatter components and was shaped to have negligible influence in the stimulus-frequency region. The stimuli employed by these authors were octave-band noise signals and they used a two-interval forced-choice technique to measure thresholds for temporal gaps. At the highest levels employed (85 dB SL) they found average thresholds of 9.17 ms for 400 - 800 Hz, 6.97 ms for 800 - 1600 Hz and 5.09 ms for 2000 - 4000 Hz. At 600 Hz the critical bandwidth is approximately 100 Hz, the time constant τ_o according to eq. (3) would be 20 ms (k set equal to 2). The value found is about one half of this, we might tentatively conclude that the auditory system would be capable of detecting a 4 dB decrement during gaps of the order of 10 ms.

At higher central frequencies the minimum detectable gap width did not decrease in proportion with the inverse of the frequency so that at higher frequencies the gap limit is not set solely by properties of the auditory filter. The main point here is that filtering effects are unmistakably present at low frequencies and that detection of a gap appears to be associated with the detection of a decrement of about 4 dB (or less at higher frequencies where k is larger). This conclusion lends support to the reasoning employed by Plomp and exemplified by Fig. 2 of this paper: the just-detectable level difference ΔS would be of the order of 3 - 4 dB. The smallest time over which this should be possible is approx. 5 - 10 ms.

A more complete study of frequency effects in gap detection was presented by Fitzgibbons (1983). He used the same technique of masking extraneous components but used wide-band as well as narrow-band maskers, the latter centred at the four frequencies 600, 1200, 2400 and 4800 Hz. For white noise he found the minimum detectable gap width to be approx. 3 ms for a sensation level of 40 dB. The main contribution was from components above 5.66 kHz since the same value was found for high-pass noise with this frequency as its cut-off frequency. By itself the bandwidth of the stimulus had little effect on the minimum detectable gap. As expected, the (central) frequency had a larger effect. The final result was expressed in the following form

$$\Delta T = 1.88 + 800/ \Delta F \qquad (4)$$

where ΔT is the minimum detectable gap length in ms and ΔF is the bandwidth of the appropriate auditory filter in Hz. Our reasoning presented above would lead to

$$\Delta T = 750/ \Delta F \qquad (5)$$

if we assume our k to be 3 and take one fourth of the time constant. A remarkably good agreement for the component due to ringing of the auditory filter. For higher central frequencies ΔT is no longer inversely related to the filter bandwidth. The cross-over takes place at the frequency where the two terms of eq. (4) are equal. This occurs at 2.67 kHz. We should conclude that a variation of approximately 2 dB occuring in a gap of 3.8 ms gives a noticeable contribution to auditory detection of that gap, an extension of the statement made earlier in this section.

In both studies described in the preceding (Fitzgibbons and Wightman 1982 and Fitzgibbons 1983) a considerable influence of the intensity level was found, larger than anticipated. This influence could be found, it is reasoned, because of the extreme caution with which splatter components were eliminated. It is obviously at the highest level of stimulation where we can expect the effects of the auditory filter to dominate, at lower levels the slow decay of masking contributes more. The intensity effect is also part of the explanation why in listeners with impaired hearing (sensorineural hearing loss) gap detection is invariably worse than in normally-hearing people (Fitzgibbons and Wightman 1982). One might have expected improved gap detection in the hearing-impaired listeners since it is well known that critical bands may be abnormally wide (de Boer 1959; Wightman et al. 1977; Florentine et al. 1980). Quite the contrary is found which suggests that the auditory decay function, or in different terms the recovery from masking, is adversely affected in these patients (Buus and Florentine 1984).

CONCLUDING REMARKS

Let us, with the foregoing in mind, try once more to confront the findings with one another. Consider a fictitious experiment in which a continuous masking signal is presented, e.g. a band of noise wider than the critical band at its centre frequency. To measure the masking power of this band of noise one can use a long tone of, say, 500 ms duration. If the intensity of the masking noise is, e.g., 40 dB per critical band, the threshold level of the test tone (at the central frequency) would be 40 dB or a few dB less. Presentation of the test tone at that level produces a 1 dB increment of level. For a test tone of 10 ms the threshold level would be 13 dB higher and for a test tone of 1 ms, wherever feasible, 23 dB higher. This difference, of course, is attributed to temporal integration, the auditory system is obviously not capable of detecting a 1 dB increment in a time of 10 ms or so. Let us now assume that the masking signal has a gap of 10 ms. This gap can be traced with a short probe tone (5 or 10 ms long) and it is found to result in a decrement of about 2 dB, provided the central frequency is sufficiently high. Under the same proviso the presence of the 10 ms gap can be detected by itself (without a probe tone) so we must assume that a decrement of 2 dB over 10 ms can be detected by the auditory system. Arguments based on the ringing of auditory filters would lead to a value of 2 - 4 dB, clearly in the correct order of magnitude. On the other hand, an increment during 10 ms can only be detected when it is approx. 13 dB.

Is this a manifestation of an extreme nonlinearity of the auditory system which makes decrements far more easy to detect than increments? Note that in physiological experiments there is a clear difference between increments and decrements (Smith and Brachman 1982) but it works just the other way around: responses to decrements reflect adaption and are thus smaller than responses to increments. The notion of an extreme asymmetry with respect to increments and decrements is not in line with current concepts about what goes on in the detection process. Detection, especially around the threshold level, is the result of a statistical decision process; if variations as small as 1 or 2 dB are involved, the decision is made on the basis of samples that have a fairly narrow distribution. By a mechanism operating on such a basis a variation of 13 dB would certainly be detected and here we are back to the same argument as the one discussed near the end of section "The auditory integrator"! No matter how one puts it, there is a clear incompatibility between models accounting for forward masking and gap detection on the one hand and the intensity integration model on the other. Moreover, thresholds of modulation cannot be explained by the integration model.

A solution to this paradox is not easy to give. If one thinks of detection as involving a moving time window, of perhaps 10 - 50 ms duration (cf. Green 1973; Penner 1975), then the threshold of gap detection should correspond to an audible increment of 2 - 4 dB at fairly small durations, for example 10 ms. Detection of tones of longer durations then should be either the same, better (if integration occurs) or worse (as a result of "forgetting"). This does not at all tally with the data. On the other hand, if detection of tone bursts during the roving window were based on fairly bad statistics (so that at 10 ms a difference of 13 dB could just be detected and temporal integration would account for better thresholds at longer durations), the data on forward masking and, particularly, gap detection could not be explained.

We should leave the problem as it is. Perhaps more comprehensive models are needed to explain all phenomena but as far as this author is aware, such a model has not been presented yet. The help of other readers is requested to put things right - if this is possible, at present. In presenting and analyzing this material the author hopes to have given food to fruitful discussions.

ACKNOWLEDGEMENTS

The author is indebted to the organizers of the 11th Danavox symposium, Drs. A. Michelsen and O. Larsen, and to the Danavox Committee for providing the means of having thorough discussions on temporal effects in hearing. Dr. P. Kuyper of this laboratory contributed to the thoughts expressed in this paper in discussions on temporal phenomena, and the author expresses his gratitude. The Netherlands Organization for Pure Research (ZWO, the Hague) was helpful in providing the means to do experiments designed to complement the present work with own experimental results (not yet reported).

REFERENCES

de Boer E (1959) Measurement of the critical bandwidth in cases of perception deafness. In: Proceedings 3rd International Congress of Acoustics, 1. Elsevier, Amsterdam, pp 100-102

de Boer E (1966) Intensity discrimination of fluctuating signals. J Acoust Soc Am 40:552-560

de Boer E (1975) Synthetic whole-nerve action potentials for the cat. J Acoust Soc Am 58:1030-1045

de Boer E (1976) On the "residue" and auditory pitch perception. In: Handbook of sensory physiology, part V/3, (eds. Keidel WD and Neff WD, Springer-Verlag, pp 469-583

de Boer E (1979) Travelling waves and cochlear resonance. In: Models of the auditory system and related signal processing techniques (eds. Hoke M and de Boer E). Scand Audiol Suppl 8, pp 17-33

Bos CE, de Boer E (1966) Masking and discrimination. J Acoust Soc Am 39:708-715

Buus S, Florentine M (1984) Gap detection in normal and impaired listeners: the effect of level and frequency. (This volume)

Duifhuis H (1973) Consequences of peripheral frequency selectivity for nonsimultaneous masking. J Acoust Soc Am 54:1471-1488

Evans EF (1975) Normal and abnormal functioning of the cochlear nerve. In: Sound reception in mammals (eds. Bench RJ, Pye A, Pye JD). Symp Zool Soc Lond nr. 37, Academic Press London, pp 133-165

Fastl H (1976) Temporal masking effects: I. Broad band noise masker. Acustica 35:287-302

Fastl H (1977) Roughness and temporal masking patterns of sinus-oidally amplitude modulated broadband noise. In: Psychophysics and physiology of hearing (eds. Evans EF and Wilson JP). Academic Press, London, pp 403-414

Feldtkeller R, Zwicker E (1956) Das Ohr als Nachrichtenempfänger. Hirzel, Stuttgart (second printing: 1967)

Fitzgibbons PJ (1983) Temporal gap detection in noise as a function of frequency, bandwidth, and level. J Acoust Soc Am 74:67-72

Fitzgibbons PJ, Wightman FL (1982) Gap detection in normal and hearing-impaired listeners. J Acoust Soc Am 72:761-765

Fletcher H (1940) Auditory patterns. Rev Mod Physics 12:47-65

Florentine M, Buus S, Scharf B, Zwicker E (1980) Frequency selectively in hearing-impaired observers. J Speech Hearing Res 23:646-669

Green DM (1973) Minimum integration time. In: Basic mechanisms in hearing (ed Møller AR). Academic Press, New York, pp 829-843

Harrison RV, Evans EF (1977) The effects of hair cell loss (restricted to outer hair cells) on the threshold and tuning properties of cochlear fibres in the guinea pig. In: Inner ear biology (eds. Portmann M and Aran JM). Inserm, Paris, pp 105-124

Houtgast T (1974) Lateral suppression in hearing (Acad thesis), Academische Pers, Amsterdam, Fig. 5.1 p 28

Jesteadt W, Bacon SP, Lehman JR (1982) Forward masking as a function of frequency, masker level, and signal delay. J Acoust Soc Am 71:950-962

Khanna SM, Leonard DGB (1983) An interpretation of the sharp tuning of the basilar membrane mechanical response. In: Mechanics of hearing (eds. de Boer E and Viergever MA). Delft University Press, pp 177-181

O'Loughlin BJ, Moore BCJ (1981) Off-frequency listening: effects on psychoacoustical tuning curves obtained in simultaneous and forward masking. J Acoust Soc Am 69:1119-1125

Moore BCJ (1978) Psychophysical tuning curves measured in simultaneous and forward masking. J Acoust Soc Am 63:524-532

Moore BCJ, Glasberg BR (1981) Auditory filter shapes derived in simultaneous and forward masking. J Acoust Soc Am 70:1003-1014

Moore BCJ, Glasberg BR (1982) Interpreting the role of suppression in psychophysical tuning curves. J Acoust Soc Am 72:1374-1379

Patterson RD (1976) Auditory filter shapes derived with noise stimuli. J Acoust Soc Am 59:640-654

Patterson RD, Henning GB (1977) Stimulus variability and auditory filter shape. J Acoust Soc Am 62:649-664

Penner MJ (1975) Persistence and integration: two consequences of a sliding integrator. Perception and Psychophysics 18:114-120

Plomp R (1964) Rate of decay of auditory sensation. J Acoust Soc Am 36:277-282

Plomp R, Bouman MA (1959) Relation between hearing threshold and duration for tone pulses. J Acoust Soc Am 31:749-758

Rodenburg M (1977) Investigation of temporal effects with amplitude modulated signals. In: Psychophysics and physiology of hearing (eds. Evans EF and Wilson JP) Academic Press, London, pp 429-437

Siebert WM (1968) Stimulus transformation in the peripheral auditory system. In: Recognizing patterns (eds. Kolers PA and Eden M). MIT Press, Cambridge (USA), pp 104-132

Smith RL, Brachman ML (1982) Adaptation in auditory-nerve fibres: a revised model. Biol Cybern 42:107-120

Verschuure J (1981) Pulsation patterns and nonlinearity of auditory tuning II: analysis of psychophysical results. Acustica 49:296-306

Viemeister NF (1973) Temporal modulation transfer function for audition. J Acoust Soc Am 53:312(A)

Viemeister NF (1977) Temporal factors in audition: a system analysis approach. In: Psychophysics and physiology of hearing (eds. Evans EF and Wilson JP). Academic Press, London, pp 419-428

Vogten LLM (1978) Low-level pure-tone masking: a comparison of "tuning curves" obtained with simultaneous and forward masking. J Acoust Soc Am 63:1520-1527

Widin GP, Viemeister NF (1979) Intensive and temporal effects in pure-tone forward masking. J Acoust Soc Am 66:388-395

Wightman F, McGee T, Kramer M (1977) Factors influencing frequency selectivity in normal and hearing-impaired listeners. In: Psychophysics and physiology of hearing (eds. Evans EF and Wilson JP). Academic Press, London, pp 295-306

Zwicker E (1956) Die elementaren Grundlagen zur Bestimmung der Informationskapazität des Gehörs. Acustica 6:365-381

Zwicker E (1965) Temporal effects in simultaneous masking by withenoise bursts. J Acoust Soc Am 37:653-663

Zwicker E (1973) Temporal effects in psychoacoustical excitation. In: Basic mechanisms in hearing (ed Møller AR). Academic Press, New York, pp 809-824

Zwicker E (1974) On a psychoacoustic equivalent of tuning curves. In: Facts and models of hearing (eds. Zwicker E and Terhardt E). Springer-Verlag, Berlin, pp 132-140

Zwicker E, Fastl H (1972) On the development of the critical band. J Acoust Soc Am 52:699-702

Zwislocki JJ (1960) Theory of temporal auditory summation. J Acoust Soc Am 32:1046-1060

Gap Detection in Normal and Impaired Listeners: The Effect of Level and Frequency

Søren Buus[1] and Mary Florentine[2]

Northeastern University, Boston, Massachusetts 02115, USA

INTRODUCTION

Temporal summation and temporal resolution are often thought to be different aspects of the same integration process. However, a long integration time is required to optimize performance in a temporal summation task, whereas a short integration time is required to optimize performance in a temporal resolution task. Indeed, the integration times found in temporal summation experiments are typically one-to-two orders of magnitude larger than those found in temporal resolution experiments (for review, see Green, this volume, see also de Boer, this volume). It is possible to reconcile this large difference either by assuming that a compressive non-linear transformation (Divenyi and Shannon 1983) or neural adaptation (Irwin and Kemp 1976; Irwin and Purdy 1982) takes place prior to integration, but in this paper we implicitly treat the two integration processes as separate by considering only temporal resolution and the corresponding short integration time.

Temporal resolution in the auditory system is governed by at least two distinctly different processes that are assessed by different types of experiments (Scharf and Buus in press). For example, one can ask listeners to decide whether two tones of different frequencies occurred simultaneously or not. Owing to the frequency selectivity of the auditory system, the two tones will stimulate, at least in part, different auditory channels. Thus, the task may be performed by comparing the timing of events between channels tuned to different frequencies. The other type of experiment is exemplified by gap detection, in which the listeners are required to detect a brief pause in an otherwise continuous sound. Plomp (1964) suggested that detection of the gap required the auditory activity to decay to some fraction of its steady-state value (see Green, this volume). Thus, gap detection depends on the decay of activity within a channel in contrast to the simultaneity judgment which depends on a comparison between channels. That these two processes are different is illustrated by the way in which they depend on frequency. Discrimination of Huffman sequences, which depends on

---

[1]Laboratory of Psychophysics, Harvard University, 33 Kirkland Street, Cambridge, MA 02138, USA and Department of Psychology (413MU), Northeastern University, Boston, MA 02115, USA.

[2]Communication Research Laboratory (133FR), Northeastern University, Boston, MA 02115, USA

comparison between channels, is independent of frequency (Green 1973). In contrast, as discussed in this paper, gap detection is strongly dependent on frequency (see also Buus and Florentine 1982; Fitzgibbons 1979, 1983, 1984a; Fitzgibbons and Wightman 1982; Florentine and Buus 1983; Shailer and Moore 1983).

The focus of this paper is the decay of auditory activity as measured by gap detection in both normal and impaired listeners. In particular, the mechanisms responsible for and the implications of the frequency dependence of gap detection will be examined. We start by presenting a simple model for gap detection and then we discuss three experiments. The first experiment tests the model's prediction of a frequency dependence of gap detection. The second experiment demonstrates how the frequency dependence of gap detection might account for the enlarged gap detection thresholds found in most impaired listeners (Boothroyd 1973; Cudahy 1977; Fitzgibbons 1979; Fitzgibbons and Wightman 1982; Florentine and Buus 1984; Irwin et al. 1981; Irwin and Purdy 1982; Tyler et al. 1982) and in chinchillas with noise induced hearing-loss (Giraudi-Perry et al. 1982). The third experiment examines this hypothesis in greater detail by comparing gap detection thresholds for listeners with real and simulated impairments. By simulated impairments we merely mean normal listeners whose pure-tone thresholds were elevated by a continuous, spectrally shaped masker to yield an audiogram similar to that of an impaired listener.

THE MODEL

One of the most fundamental properties of the auditory system is its ability to separate spectrally disparate sounds. This frequency analysis can be likened to the operation of a bank of bandpass filters (de Boer 1959). If an incoming sound is suddenly switched off, one might expect that the output of the auditory filter, like that of any physical filter, does not stop immediately, but decreases toward zero over some time. To a first approximation, the time constant for the decay of this ringing can be assumed to be inversely proportional to the bandwidth of the filter. Because the auditory filter is wider at high than at low frequencies, auditory activity ought to decay faster and temporal resolution improve as frequency increases. This simple consideration is the point of departure for our model of gap detection.

The role of the auditory filter in the decay of auditory activity as measured by forward masking was investigated in considerable detail by Duifhuis (1972, 1973). He used a model consisting of a filter and a stochastic spike generator to explain several phenomena related to forward masking. In the following, we shall investigate a somewhat simpler model for gap detection. The model does not include several adaptation and suppression effects that have been observed in physiological experiments and that probably are important for gap detection. As stated by Evans (this volume), the comprehensive data necessary for modeling are not yet available. As shown in Fig. 1, the model is basically an energy-detector with a filter at the input. It is similar to that used by Buunen

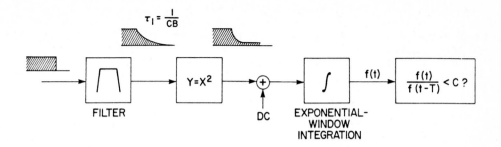

Fig. 1. A simplified model of temporal gap detection. The output of a filter, whose ringing decays with a time constant, τ_1, equal to the inverse of the critical bandwidth (1/CB), is converted into instantaneous power. A small DC component is added to account for absolute threshold. The resulting function is integrated and led to a very simple energy detector. Detection occurs when the energy output decreases a fixed proportion of the steady-state output that was present before the gap. The short-term variability in the energy is taken into account by making the criterion, C, depend on the bandwidth of the filter and the integration time of the exponential integrator.

and van Valkenburg (1979) to explain their data on gap detection in wide-band noise, except that they did not include the ringing of the auditory filter in their calculations. In our model, we have assumed that the ringing of the filter has an exponential decay with a time constant equal to the inverse of the critical bandwidth tabulated by Zwicker (1961, see also Zwicker and Feldtkeller 1967; Scharf 1970). The output of the filter is squared to yield the instantaneous power of the filtered input. To account for absolute threshold, a small DC component is added to the instantaneous power, which is then integrated with an exponential window as proposed by Munson (1947) and Zwislocki (1969).

If the rise and fall times are zero, and the DC-component that accounts for threshold is ignored, a simple formula can be derived for the behavior of this model. According to Green (this volume), the sensitivity, d', of an ideal energy detector is

$$d' = \sqrt{W\tau} \cdot \frac{y(0) - y(t)}{y(0)}$$

where W is the bandwidth of the filter, τ is the integration time, y(0) is the average output of the integrator at the beginning of the gap (which is equal to the steady-state average), and y(t) is the average output of the integrator at the end of a gap with duration t. The factor $\sqrt{W\tau}$ describes the variability in the output of the integrator. If an additional internal noise is assumed to corrupt the ideal performance of the detector this factor becomes

$$\sqrt{\frac{W\tau}{W\tau\sigma_I^2 + 1}}$$

where σ_I is the relative standard deviation of the internal noise.

To derive $(y(0) - y(t))/y(0)$ we first consider the normalized output of the filter, $G(t)$:

$$G(t) = g(t)/g(0) = \begin{cases} 1 & t \stackrel{<}{=} 0 \\ \exp(-Wt) & t > 0 \end{cases}$$

where $g(0)$ is the average output of the filter at the beginning of the gap (which is equal to the steady-state average), $g(t)$ is the average output of the filter at the end of a gap with duration t, and W is the bandwith of the filter. (The time constant τ_1, in Fig. 1 is equal to $1/W$). The normalized output of the square-law rectifier is the instantaneous power, $H(t)$:

$$H(t) = G^2(t) = \begin{cases} 1 & t \leq 0 \\ \exp(-2Wt) & t > 0 \end{cases}$$

Finally, the normalized output of the integrator, $Y(t)$, is

$$Y(t) = y(t)/y(0) = \int_{-\infty}^{t} H(T) \exp(-(t-T)/\tau)\, dT$$

where τ is the time constant for the integrator.
Solving this integral and reducing the expression yields

$$\frac{y(0)-y(t)}{y(0)} = 1 - Y(t) = 1 - \frac{2W\tau}{2W\tau-1}\exp(-t/\tau) + \frac{1}{2W\tau-1}\exp(-2Wt)$$

and thus,

$$d' = \sqrt{\frac{W\tau}{W\tau\sigma_I^2+1}} \cdot \left[1 - \frac{2W\tau}{2W\tau -1}\exp(-t/\tau) + \frac{1}{2W\tau-1}\exp(-2Wt) \right]$$

where W is the bandwidth of the initial filter, τ is the time constant for the integrator, σ_I is the relative standard deviation of the internal noise, and t is the duration of the gap. Notice that the formula describes only the asymptotic performance at high levels and does not depend on the overall level of the stimulus. Moreover, this formula is only valid for $W\tau \gg 1$ (see Green, this volume).

To arrive at predictions for our experiment, however, we used a numerical model which permitted us to take the rise/fall time, the noise floor in the signal band, and the signal level relative to detection threshold into account. To keep the calculations simple we assume that a change in the short-term energy is just detectable when it exceeds some fraction of its steady-state value, as illustrated in Fig. 1. The criterion depends on the variance of an assumed internal noise and the variance of the steady-state output of the short-term integrator, which in turn depends on the bandwidth of the initial filter and the time constant of the integrator. We also assumed that three identical, independent channels were working in parallel, such that the criterion was set to correspond to a decrease equal to 0.58 $(1/\sqrt{3})$ times that derived for a single channel. The criterion is independent of level because the internal noise is assumed to be proportional to the long-term average of the energy, i.e. the detector obeys Weber's Law.

The proportionality constant for the internal noise and the time constant for the integrator are the only free parameters in the model. Our calculations show that these two parameters to a large extent can be traded off against one another so the predictions of the model are not critically dependent on the assumed time constant. The predictions, which will be discussed later, were obtained with a time constant of 30 ms and a proportionality constant for the internal noise set to render a level decrement of 0.7 dB just detectable for a deterministic signal within a single channel (critical band). Except at low levels, where those shown fit the data slightly better, similar predictions are obtained with an integration time of 3 ms and an internal noise corresponding to a single-channel DL of about 8 dB.

EXPERIMENT I

Method

The first experiment was designed to test the model's predictions of the effect of level and frequency by measuring the minimum detectable gap duration, MDG, for octave bands of noise. The center frequency was varied from 0.25 to 14 kHz and the spectrum level was varied from near threshold to 45 dB SPL. The slopes of the skirts were 18 dB/octave. To permit comparison with data in the literature, the MDG for a wide-band (low-pass, f_c = 7 kHz) noise was also measured. The gap was produced by turning off and on the otherwise continuous signal with a 1-ms fall and rise time. The gap duration is defined as the time between the beginning of the offset to the beginning of the onset. The gap duration measured between the half-power points is about 0.2 ms longer than that defined above. To ensure that the spectral splatter produced by the rapid offset and onset of the signal was inaudible, each octave band of noise was presented with its complementary band-stop masker. (No masker was used with the wide-band noise.) The spectrum level of the masker was equal to that of the signal and the slope of the filter skirts was 48 dB/octave such that the masker was about 21 dB below the signal at the center frequency. The masker was presented continuously.

The signals were produced by a noise generator (Grason-Stadler 455C) whose output was band-pass filtered (Krohn-Hite 310AB), gated (Coulbourn S84-04), and attenuated (Hewlett-Packard 350B) before it was added passively to the masker. The masker was produced by a noise generator (General Radio 1382A) whose output was band-stop filtered (Krohn-Hite 3343) and attenuated (Hewlett-Packard 350D) before it was added to the signal. The output of the summation network was attenuated (Hewlett-Packard 350D) and led to an amplifier (Scott A406) which fed the earphone through an impedance-matching network. For the wide-band noise and the center frequencies at or below 4 kHz, TDH-49 earphones were used; for the 8 and 14 kHz center frequencies, Yamaha YH-1 earphones were used.

The MDG was measured monaurally in a two-interval, two-alternative, forced-choice paradigm with feedback. Two 500-ms lights presented 500 ms apart marked the observation intervals. The listener's task was to judge which interval contained the gap, which began 50 ms after the onset of the light marking the appropriate interval. For each listener, the MDG was determined as the average of MDGs obtained in three adaptive runs using a modified BUDTIF procedure (Campbell 1963). This procedure required a brief pause after each block of four trials to permit the experimenter to set the gap duration for the next block. If all four trials were correct, the gap duration was decreased by one step of approximately 10%; if three of the four trials were correct, it remained the same; if two or fewer trials were correct it was increased by one step. Each run was terminated at the sixth reversal subsequent to 40 warm-up trials. The MDG determined in this manner converges toward the gap duration required for 73.4% correct responses, corresponding to a d' of about 0.88. Prior to data collection, each listener was given practice until no systematic change in performance was apparent.

Three young college students with normal hearing participated in the experiment. Their audiometric thresholds were within 10 dB HL (ANSI 1969) and they had a negative medical history for hearing impairments.

Results and discussion

Figure 2 shows the MDGs plotted as a function of overall SPL. In this and the following figures, overall SPL is the total for the masker and the signal calculated for the 7-kHz bandwidth of the TDH-49 earphone. Thus, the spectrum level is always 38 dB less than the overall level indicated in the figures. As shown in Fig. 2, the MDG for the wide-band noise decreases with increasing level up to about 40 dB SPL, reaches a minimum at 60 to 70 dB SPL, above which it increases slightly. These data are in excellent agreement with other data for gap detection in a wide-band noise (Florentine and Buus 1982, 1984; Irwin et al. 1981; Penner 1977; Plomp 1964). Although small, we believe that the increase in the MDGs above 70 dB SPL is real, because it is apparent in Irwin et al.'s (1981), Plomp's (1964), and Penner's (1977) data as well as in our individual data.

As expected, the data for the octave bands of noise show a pronounced effect of frequency. For example, for the 25-dB spectrum level, the MDG decreases by a factor of 10, from about 54 ms to about 5 ms, as the center frequency increases from 0.25 to 8 kHz. These data are similar in form to the data published by Fitzgibbons (1983, 1984a) and Shailer and Moore (1983), but our MDGs are about double those obtained by Fitzgibbons and about 30% larger than those obtained by Shailer and Moore. The difference is most likely the result of the different relative levels of the band-stop masker used in the different studies. The difference between the spectrum

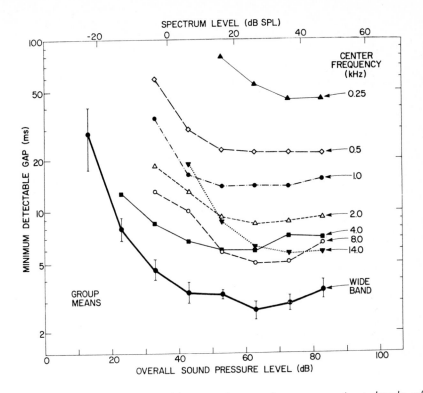

Fig. 2. The minimum detectable gap for various one-octave bands of noise and wide-band noise is plotted as a function of overall SPL on the bottom abscissa and spectrum level on the top abscissa. The center frequency is given for the one-octave noise bands. Each point is the geometric mean of data obtained from three normal listeners. The vertical bars show plus and minus one standard deviation. Standard deviations are omitted from the octave-band data for the sake of the clarity; they are approximately the same as for the wide-band condition.

levels of the band-stop masker and the band-pass signal was 20 dB in Fitzgibbons' study, 10 dB in Shailer and Moore's study, and 0 dB in the present study.

Some might argue that it is not the center frequency and change in auditory-filter bandwidth that are the important variables in our experiment. Owing to the one-octave bandwidth, the absolute band-width in Hz was proportional to the center frequency. Thus, it could be the relevant variable. Indeed, the formula derived by Green (this volume) for the ideal energy detector suggests that d' should be proportional to the square-root of the signal bandwidth. Moreover, the four-fold decrease of the MDGs from 0.5 to 8 kHz is exactly predicted by the square-root dependence. However, the change in the MDGs between other pairs of center frequencies are less accurately predicted. More importantly, Shailer and Moore (1983) found only a small effect of bandwidth when the center frequency was kept constant, and no more than expected on the basis of the slight increase in upper cut-off frequency with increasing bandwidth. Although we must admit that our data do not permit a clear distinction between the signal-bandwidth and the center-frequency hypotheses, we believe that center frequency and the bandwidth of the auditory filters are the relevant variables. The following discussion rests on this assumption.

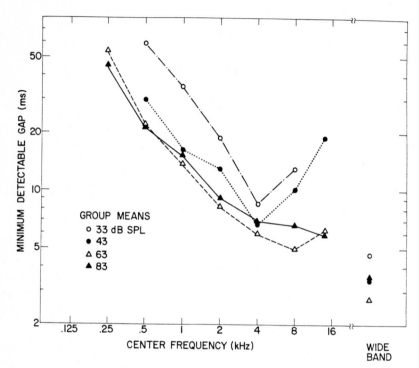

Fig. 3. The minimum detectable gap for octave bands of noise is plotted as a function of the center frequency of the noise at four different test levels. For comparison, the data for wide-band noise are also plotted. Each point is replotted from Fig. 2.

The effect of frequency is clearly shown in Fig. 3 which depicts the MDG as a function of frequency with overall SPL as the parameter. The data are the same as in Fig. 2. At 33 and 43 dB, the MDG is nearly inversely proportional to center frequency up to 4 kHz. Above 4 kHz, the MDG increases. At 63 dB SPL, the MDG decreases up to 8 kHz, but above approximately 1 kHz, the decrease is slower than inversely proportional to the frequency. At 83 dB SPL, the MDG decreases monotonically with increasing frequency, but the decrease is slower than inversely proportional to the frequency above approximately 0.5 kHz and very slow above 4 kHz. The increase above 4 or 8 kHz seen at the low-to-moderate levels is probably due to the normal increase in threshold toward the high frequencies, so that the signal was either inaudible or at a very low SL.

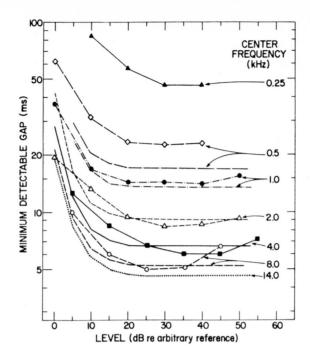

Fig. 4. Model predictions for the minimum detectable gap are compared to the data. The data from Figure 2 are plotted as a function of level for one-octave noise bands at different center frequencies. Because the reference level in the model is arbitrary, the data have been shifted along the X-axis by eye to yield the best fit to the predictions. We chose to shift the data rather than the predictions for the sake of clarity.

As seen in Fig. 4, our simple model predicts correctly the major effects in the data, but not all the details. On the positive side, the model predicts that the MDG should decrease for the first 15 or 20 dB after which it should stay constant, in good agreement with the data. Furthermore, it also predicts that the MDG should decrease with increasing frequency, and the magnitude of the decrease is in reasonable agreement with the data between 1 and 8 kHz. On the negative side, the model predicts that the MDG should be lower at 14 than at 8 kHz, which it is not. It also predicts that the increase in the MDG for frequencies below 1 kHz should be substantially smaller than that shown by the data. The predicted MDG is about 25% below the data at 0.5 kHz and about 60% at 0.25 kHz. (Because the critical bandwidth is about the same at 0.25 and 0.5 kHz the model predicts that the MDGs should be about the same at these two frequencies). However, the difference between the predictions and the data at the extreme frequencies may be due to the rapid increase in absolute threshold at these frequencies which renders the effective bandwidth of the signal less than one octave. Furthermore, below 0.5 kHz newer estimates of the auditory-filter bandwidth are not constant, but continue to decrease (e.g. Moore and Glasberg 1983). If these estimates were used for the model, it would predict larger MDGs at 0.25 than at 0.5 kHz. Some details of the predicted effect of level are also problematic. The predicted increase of the MDG at low levels appears more severe than that indicated by the data. Moreover, the MDG is not constant at high levels as predicted by the model, but increases slightly above 60 or 70 dB SPL.

Finally, in its current formulation the model does not predict the MDGs for a wide-band noise. It cannot explain why the MDG for a wide-band noise is substantially lower than the lowest MDG obtained for any of the octave bands. In our opinion, this is probably the result of several channels working in concert, such that information is combined across frequency to improve detection of the gap. If the model is reformulated to consist of a number of frequency selective channels from which information is combined -- perhaps according to an optimum decision rule -- it would predict better performance for a wide-band signal than for any individual band. In fact, a rough prediction might be that, at high levels, the wide-band MDGs should be about 1.7 ($\sqrt{3}$) times lower than the high-frequency MDGs, because the MDGs are about the same for the three octave bands at and above 4 kHz, but larger at lower frequencies. This rough prediction is in agreement with the data. Presumably, a multi-band model could also account for the increased MDGs at the extreme frequencies, because the smaller effective bandwidth of the signal makes fewer channels useful for detection of the gap.

Despite its shortcomings, we find the model useful in interpreting the data. It suggests that the auditory filter plays an important role in determining the temporal resolution of the auditory system, at least for frequencies below 4 kHz. It also suggests that the increase of the MDG at low levels results from the absolute threshold effectively limiting the on/off ratio at low levels. That most of the decrease in the predicted MDGs takes place over the first 10-15 dB is in agreement with the data of Penner (1975) and of Irwin and Purdy (1982) showing that partially filling the gap with a continuous noise only affects the MDG when the on/off ratio is less than 10-15 dB.

The model may also help us evaluate the effect of some of the non-ideal aspects of the physical stimulus. First, to contain the spectral splatter within reasonable limits, the gap had fall and rise times of 1 ms. If the decay of auditory activity is fast, the fall and rise times may limit the performance. Another non-ideal aspect of our stimuli is the finite rejection of the noise within the stop-band of the continuous band-stop masker. Although the rejection was about 21 dB at the center frequency, the average rejection across the one-octave signal bandwidth was only about 10 dB. Both non-ideal aspects of the stimulus were included in the predictions shown in Fig. 4, but to evaluate their effect we tried to omit them as shown in Fig. 5. The predicted asymptotic MDGs for four stimulus conditions with actual or ideal rise/fall times and actual or ideal stop-band rejection ratio are plotted as a function of frequency. For comparison, the data for 63 dB SPL are also plotted. The long-dashed line, which is the prediction for the actual stimuli, is in quite good agreement with the data. Comparison of

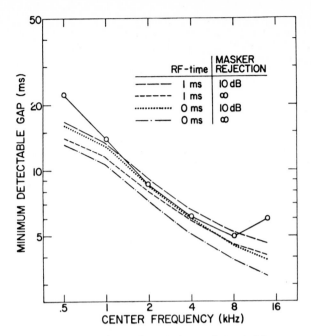

Fig. 5. Model predictions for the minimum detectable gap are used to assess how the 1-ms rise/fall time and the presence of masker energy in the signal band might affect the measured MDGs. The 10-dB masker rejection assumed in the model predictions represents an average across the one-octave signal bandwidth. For comparison, the data for a spectrum level of 25 dB SPL are shown by the circles.

the long-dashed and the dotted lines gives the predicted effect of the 1-ms decay time of the signal. On the logarithmic plot, the difference is very small at low frequencies and increases toward higher frequencies. At all frequencies, the difference is a little less than 1 ms. Comparison of the long-dashed and short-dashed lines gives the predicted effect of the finite stop-band rejection. The difference is 15% and nearly independent of frequency. Comparison of the long-dashed and dash-dotted lines indicates that our measured MDGs may be 25–30% higher than what might have been obtained if it were possible to have infinitely sharp filter skirts and to have instantaneous fall and rise times without introducing audible spectral splatter. Most importantly, however, the non-ideal aspects of our stimuli probably did not change the effect of frequency; the predicted functions for the actual and the ideal stimuli are largely parallel.

EXPERIMENT II

Experiment I clearly demonstrates that gap detection is better at high than at low frequencies. Thus, one might expect that the MDGs for a wide-band noise should depend on the extent to which the high frequencies are audible. Listeners who cannot hear the high frequencies due to filtering, masking, or hearing loss ought to show larger MDGs than listeners who can hear the entire frequency range. It is possible, then, that impaired listeners show larger-than-normal MDGs (Boothroyd 1973; Cudahy 1977; Fitzgibbons 1979; Fitzgibbons and Wightman 1982; Florentine and Buus 1984; Irwin et al. 1981; Tyler et al. 1982) not because they have reduced temporal resolution, <u>per se</u>, but because they simply cannot hear the frequencies that yield the best gap detection. Experiment II provides a preliminary test of this hypothesis by comparing gap detection for low-pass, high-pass, and wide-band noises.

Method

Gap detection was measured as a function of level in three young college students using the same procedure as in Exp. 1. The low-pass and high-pass signals were presented with complementary maskers, which were continuous and had the same spectrum level as the signal. The cut-off frequency for the low-pass and high-pass noises was 2 kHz. This frequency was chosen to make the low-pass noise encompass the range of frequencies that might be audible to a listener with a typical high-frequency hearing loss.

Results and discussion

As shown in Fig. 6, the results for the three listeners are very similar as indicated by the very small standard deviations, except at the lowest levels. The MDGs for low-pass noise are two-to-five times larger than those for either high-pass or wide-band noise. As expected, limiting the gap to occur only at frequencies below 2 kHz severely impairs its detection. On the other hand, the MDGs for high-pass and wide-band noise are nearly the same, except perhaps at overall SPLs less than approximately 45 dB. This indicates that frequencies below 2 kHz do not provide much information for gap detection in a wide-band noise.

The extent to which the inaudibility of high frequencies may account for the enlarged MDGs observed in some impaired listeners is illustrated by comparing our results for low-pass noise to those for impaired listeners. The crosses in Fig. 6 show the average MDGs for seven of Irwin et al.'s (1981) impaired listeners, who had bilateral high-frequency hearing losses of predominantly sensorineural origin. The impaired listeners' MDGs in wide-band noise are only slightly higher than our normal listeners' MDGs for low-pass noise. It seems clear that the elevated high-frequency thresholds bear the primary responsibility for the enlarged MDGs. However, the difference between the normal low-pass MDGs and the impaired wide-band MDGs may indicate reduced temporal resolution <u>per se</u>, in the impaired listeners. The next experiment was designed to separate more clearly the effect of elevated thresholds from the effect of a possible reduction in temporal resolution.

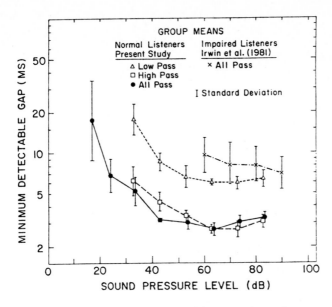

Fig. 6. The average minimum detectable gap for three normal listeners, plotted as a function of overall SPL, is compared with the average from seven impaired listeners. Triangles, squares, and circles show geometric means of normal listeners' data for low-pass, high-pass, and all-pass noise, respectively. The crosses show the geometric means of data obtained by Irwin et al. (1981) for seven listeners with high-frequency sensorineural losses. The vertical bars show plus and minus one standard deviation.

EXPERIMENT III

To separate the role of the audiometric configuration from that of a possible reduction in temporal resolution, we measured the MDGs for three groups: listeners with normal hearing, listeners with impairments of primarily cochlear origin, and normal listeners with simulated impairments produced by a spectrally shaped masking noise. Comparing the performance of real and simulated impairments is a powerful technique to separate the effects of impaired listeners' elevated thresholds from the effects of other alterations in their impaired auditory system. Because the thresholds are the same for a real impairment and its simulation, a test performed at the same SPL will also be at the same sensation level, SL, in the two groups. Moreover, masking produces the reduced dynamic range and the loudness recruitment that are typical of cochlear impairments (Steinberg and Gardner 1937). On the other hand, masking does not affect frequency selectivity (Green et al. 1981), temporal summation (Zwicker and Wright 1963), loudness summation across frequency (Scharf and Hellman 1966; Hellman and Scharf 1984), and probably not temporal resolution, per se (cf. Fitzgibbons 1984b). Accordingly, listeners with simulated impairments ought to show abnormal results only to the extent that the performance is affected by the elevated thresholds and abnormal intensity perception. Differences between simulated and real impairments must be ascribed to abnormalities in frequency selectivity and/or temporal processing caused by the cochlear impairment. In the present experiment, comparisons between simulated and real impairments are used to investigate impaired listeners' temporal resolution as measured in gap detection.

Method

The MDG for a wide-band noise was measured as a function of level using the same procedure as in Exp. I. The stimulus was a white noise which was low-pass filtered (18 dB/octave) with a 7-kHz cut-off frequency. The noise (General Radio 1382 or Grason-Stadler 455C) used to elevate the thresholds for the simulated impairments was spectrally shaped (Crown EQ-2, two channels in series), filtered (Krohn-Hite 3343), and attenuated (Hewlett-Packard 350D) before being summed with the signal which was produced by the apparatus described in Exp. I.

Six normal listeners, 20 to 50 years old, four impaired listeners, 20 to 36 years old, and four listeners with simulated impairments (two listeners for each of two audiograms) were tested. The normal listeners had audiometric thresholds within 10 dB HL (ANSI 1969) between 0.25 and 8 kHz and a negative medical history for hearing impairments. The audiometric thresholds for the impaired listeners are shown in the top panel of Fig. 7, which also shows their thresholds at 10, 12, 14, 16, and 18 kHz plotted relative to the average threshold for 17 young (18 to 32 years), normal listeners. (Missing points indicate that the threshold was beyond the limits of the apparatus.) The impairments were of primarily cochlear origin as evidenced by the results of an extensive audiological examination. The audiometric thresholds for the simulated impairments were within 3 dB of those for the loss being simulated. (More information about this experiment as well as data for additional real and simulated impairments can be found in Florentine and Buus (1984)).

Results and discussion

The MDGs for all listeners are shown in the bottom panel of Fig. 7. The MDGs for the six normal listeners decrease from 25 ms at 20 dB SPL to about 3 ms at 50 dB and above. These data are in good agreement with previous data (Buus and Florentine 1982; Fitzgibbons 1983, 1984a, 1984b; Florentine and Buus 1982, 1983; Irwin et al. 1981; Penner 1977; Plomp 1964). The small standard deviations indicate that the MDGs differ relatively little among normal listeners, except at low levels. Accordingly, one would expect the comparison between a real impairment in one listener and a simulated impairment in another not to differ significantly from the more desirable comparison of the impaired ear to the masked normal ear in the same unilaterally impaired listener. In fact, data from a unilaterally impaired listener support this hypothesis (Florentine and Buus 1984).

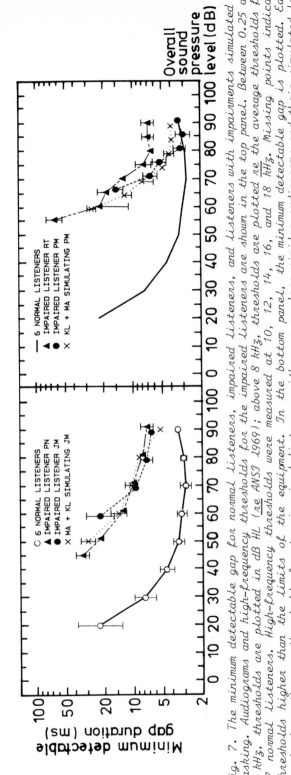

Fig. 7. The minimum detectable gap for normal listeners, impaired listeners, and listeners with impairments simulated by masking. Audiograms and high-frequency thresholds for the impaired listeners are shown in the top panel. Between 0.25 and 8 kHz, thresholds are plotted in dB HL (re ANSI 1969); above 8 kHz, thresholds are plotted re the average thresholds for 17 normal listeners. High-frequency thresholds were measured at 10, 12, 14, 16, and 18 kHz. Missing points indicate thresholds higher than the limits of the equipment. In the bottom panel, the minimum detectable gap is plotted. Each graph shows the gap thresholds for two impaired listeners with more or less similar audiograms and their simulated-loss counterpart. For comparison, gap thresholds for unmasked normal listeners are also shown. Except for the normal listeners, the vertical bars show the range encompassed by plus and/or minus one standard deviation of three adaptive runs. For the normal listeners, the vertical bars show standard deviations calculated across six listeners. When the standard deviation is less than the size of the point, only the point is shown.

The results for JM, aged 36, and PN, aged 22, who have steeply sloping audiograms and severe high-frequency losses are shown in the left graphs. JM's thresholds above 10 kHz are beyond the limits of the apparatus and PN's thresholds also indicate a substantial loss above 8 kHz. The tendency for the loss to decrease as frequency increases beyond 10 kHz results mostly from the increase of the average normal thresholds towards higher frequencies and should not be taken to indicate that PN might use information at the ultra-audiometric frequencies for gap detection. The thresholds, even in normal listeners, are so high that frequencies above 14 kHz are unlikely to contribute information.

The MDGs for JM and PN are quite similar and show enlarged MDGs at all levels. The MDGs for the two listeners simulating JM's loss are very similar to those for the real losses, and all three functions are similar to those for Irwin et al.'s (1981) impaired listeners with sloping high-frequency losses (see Fig. 6). The similarity between real and simulated losses indicates that elevated thresholds are responsible for the impaired listeners' enlarged MDGs.

The results for PM, aged 22, and RT, aged 20, are shown in the right graphs. Their mildly sloping audiograms are quite similar and differ no more than 15 dB at any frequency up to 18 kHz. (Although their performance on the audiometric test battery was quite similar, their hearing losses probably have different etiologies. Their medical histories indicate that RT's hearing loss probably is hereditary, whereas PM's hearing loss appears to be related to childhood disease and may be drug-induced.) As similar as these listeners are with respect to audiogram, age, and educational background (both were college students), they have very different gap functions. The MDGs for PM are enlarged at low levels, but normal above 80 dB. In contrast, the MDGs for RT are enlarged at all levels. The MDGs for the listeners simulating PM's loss are nearly identical to those for PM, and above 80 dB, both are about 1.5 to 2 times lower than those for RT. The difference between RT and the simulated and real losses for PM is not due to differences in ultra-audiometric thresholds. RT's thresholds are slightly lower than PM's above 8 kHz, and simulating PM's loss up to 20 kHz yielded results almost identical to those shown in the figure. Thus, the enlarged MDGs at high levels probably indicate that RT truly has reduced temporal resolution.

Taken together, the comparisons between our real and simulated losses clearly indicate that elevated pure-tone thresholds bear the prime responsibility for all four impaired listeners' enlarged MDGs at all but the highest level. In fact, the results for three impaired listeners can be accounted for entirely by their elevated pure-tone thresholds as evidenced by the similarity between the simulated and real losses over the entire dynamic range. In their data for seven impaired listeners (including the four discussed here), Florentine and Buus (1984) found this to be true for four listeners, whereas three listeners appeared to have reduced temporal resolution as evidenced by smaller MDGs in simulated than in real losses at the highest levels.

The finding that truly reduced temporal resolution is observable only at the highest levels indicates that a gap detection test should be performed above 85 dB SPL such that the signal is well above threshold for the impaired listener. Fitzgibbons (1984b) also advocated comparing gap detection in normal and impaired listeners at equal SPL with the signal at least 20 dB above threshold. Even then, however, our data for listeners with steeply sloping, severe high-frequency losses indicate that an enlarged MDG does not necessarily indicate reduced temporal resolution, per se. As one should expect on the basis of the frequency dependence of the normal temporal resolution, it matters which part of the signal accounts for its audibility. If the important high-frequency part is inaudible, whether due to hearing loss or masking, the MDGs will be larger than for unmasked normal listeners. Thus, the interpretation of impaired listeners' MDGs for a wide-band noise must take into account the configuration of the hearing loss and the effect of elevated thresholds. Nevertheless, when the hearing loss at 4 and 8 kHz is below approximately 60 dB, it seems that abnormally large MDGs at the high levels indicate reduced temporal resolution.

The data from the impaired listeners also have some interesting implications when interpreted in light of our simple model. Because the ringing of the initial filter decreases with increasing bandwidth, one might expect that impaired listeners with broader tuning curves (e.g., Florentine et al. 1980) should show better gap detection than normal listeners. However, at high frequencies, which are the most important for gap detection, the auditory filters are so wide, even in normal listeners, that the time constant for the ringing is considerably less than the time constant for the short-term integrator and little benefit would result from a widening of the filter. Thus, at high levels, when the high-frequency part of the wide-band signal is clearly audible, the MDGs primarily reflect the state of the short-term integrator.

At low levels, however, the impaired listeners presumably could not hear the high-frequency part of the signal and were forced to rely on information at lower frequencies. In this case, the ringing of the filter ought to be the primary limitation on temporal resolution and one might expect that the MDG should benefit from a possible widening of the auditory filters in the impaired listeners. The data for the simulated losses indicate no such benefit, however. The similarity of the low-level MDGs in real and simulated impairments indicates that the performance by most of the cochlearly impaired listeners can be accounted for by assuming that listening is restricted to low-frequency channels with normal temporal resolution. However, careful inspection of RT's MDGs below 80 dB SPL reveals that she may have reduced temporal resolution at low frequencies. For levels above 60 dB SPL, she consistently has larger MDGs than PM and the listeners simulating PM, although her hearing loss at low frequencies is about 10 dB less than PM's.

To further confirm that RT indeed has reduced temporal resolution, whereas the other listeners do not, the MDGs were measured for octave bands of noise centered at 0.5 kHz for RT and PN and centered at 4 kHz for RT and PM. The method, stimuli, and apparatus were the same as in Exp. I. The results are shown in Fig. 8. For comparison, the results for normal listeners, replotted from Fig. 2, are also shown. The MDGs for PN (at 0.5 kHz) and PM (at 4 kHz) are virtually identical to those for the normal listeners, except near PM's elevated threshold for the 4-kHz signal. Thus, it seems that their temporal resolution is indeed normal. One might question whether PN has normal temporal resolution at frequencies outside the range of normal pure-tone thresholds. Unfortunately, his loss is too severe to permit testing at 2 or 4 kHz, but his MDGs at 1 kHz are normal once the level is well above threshold. This indicates normal temporal resolution despite a 35-dB hearing loss. In contrast, the MDGs for RT are clearly elevated at all levels. All indications are that she has reduced temporal resolution, certainly at 0.5 and 4 kHz, and probably throughout the audible frequency range. (Further measurements at 1 kHz also indicate reduced temporal resolution.)

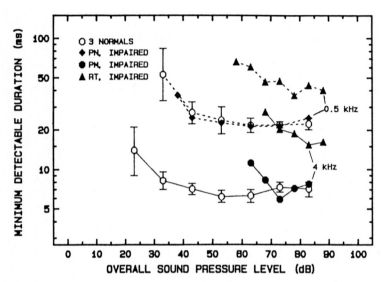

Fig. 8. The minimum detectable gap for octave bands of noise at 0.5 and 4 kHz in normal and impaired listeners. The data are connected by dashed lines for 0.5 kHz and by solid lines for 4 kHz. The data for normal listeners (open circles) are replotted from Figure 2. Vertical bars show plus and minus one standard deviation. Filled symbols show data for impaired listeners RT (triangles), PM (circles), and PN (diamonds). The comparison with simulated losses indicated that RT has reduced temporal resolution, whereas PM and PN have normal temporal resolution. These data support that conclusion.

Together, these findings confirm the inferences drawn from rather detailed comparisons between impaired listeners and their simulated-loss counterparts. These comparisons do seem to accomplish our goal, which was to separate the effects of elevated thresholds and abnormal intensity perception from the effects of other auditory deficits such as reduced spectral and/or temporal resolution. Indeed, we have been able to show that not all listeners who show enlarged MDGs have reduced temporal resolution, but some do. It is interesting to note that in the present sample of impaired listeners, as well as in the larger sample presented by Florentine and

Buus (1984), there is no clear relation between the severity of the loss and the presence of reduced temporal resolution. It is possible that reduced temporal resolution is found only in some etiologies and that different etiologies of our listeners' hearing losses are responsible for the differences in their temporal resolution, but the present study was not designed to differentiate among etiologies. Thus, we are left with a finding that is not uncommon in research with impaired listeners: some do and some don't, and we don't know why.

SUMMARY

The main findings from these and other experiments on gap detection may be summarized as follows:
1) The MDG decreases with increasing level up to 20 or 30 dB, above which it is relatively constant.
2) Data from several studies show a small, but consistent, increase of the MDG as the signal level increases above 60-70 dB SPL.
3) The MDG for bands of noise decreases with increasing frequency up to 4 or 8 kHz.
4) To a first approximation, a simple model in which a band-pass filter precedes an energy detector can account for the data for octave bands of noise. The model indicates that the auditory filter plays a major role in limiting the temporal resolution at frequencies below 4 kHz. A multi-band version of the model may also account for the wide-band data and may fit data at the extreme frequencies better than the simple single-band model investigated in this paper.
5) Owing to the frequency dependence of gap detection, the MDG in impaired listeners depends on the configuration as well as the amount of the loss. In fact, it appears that the enlarged MDGs in impaired listeners mostly result from their inability to hear the important high-frequency part of the signal.
6) Some impaired listeners show worse MDGs at high levels than their simulated-loss counterparts, indicating that elevated thresholds are not sufficient to explain their enlarged MDGs. Presumably, these listeners have reduced temporal resolution.

ACKNOWLEDGMENTS

We thank Professor David Green and Ms. Rhona Hellman for reading an earlier version of this manuscript and Professor John Irwin for providing us with the raw data for the impaired listeners in Fig. 6. We also would like to thank Ms. Sharon Quigley for help in preparing the manuscript. Part of this research was supported by National Institute of Health grants RO1NS18280 and RRO7143.

REFERENCES

ANSI (1969) American National Standards Institute. Specifications for Audiometers. S3.6-1969. American National Standards Institute, New York

Boer E de (1959) Measurement of the critical bandwidth in cases of perception deafness. Proceedings 3rd International Congress of Acoustics. 1, Elsevier, Amsterdam pp 100-102

Boothroyd A (1973) Detection of temporal gaps by deaf and hearing subjects. (SARP NO. 12). Clarke School for the Deaf, Northampton.

Buunen TJF, van Valkenburg DA (1979) Auditory detection of a single gap in noise. J Acoust Soc Amer 65:534-537

Buus S, Florentine M (1982) Detection of a temporal gap as a function of level and frequency. J Acoust Soc Amer 72 (Suppl. 1), S89

Campbell RA (1963) Detection of a noise signal of varying duration. J Acoust Soc Amer 35:1732-1737

Cudahy EA (1977) Gap detection in normal and hearing-impaired listeners. Paper presented to the American Speech and Hearing Association, Chicago

Divenyi P, Shannon RV (1983) Auditory time constants unified. J Acoust Soc Amer 74 (Suppl. 1) S10

Duifhuis H (1972) Perceptual analysis of sound. Doctoral Dissertation, University of Technology, Eindhoven, Holland

Duifhuis H (1973) Consequences of peripheral frequency selectivity for nonsimultaneous masking. J Acoust Soc Amer 54:1471-1488

Fitzgibbons P (1979) Temporal resolution in normal and hearing-impaired listeners. Doctoral Dissertation, Northwestern University, Evanston, IL.

Fitzgibbons P (1983) Temporal gap detection in noise as a function of frequency, bandwidth, and level. J Acoust Soc Amer 74:67-72

Fitzgibbons P (1984a) Tracking a temporal gap in band-limited noise: Frequency and level effects. Perception & Psychophysics 35:446-450

Fitzgibbons P (1984b) Temporal gap resolution in masked normal ears as a function of masker level. J Acoust Soc Amer 76:67-70

Fitzgibbons P, Wightman FL (1982) Gap detection in normal and hearing-impaired listeners. J Acoust Soc Amer 72:761-765

Florentine M, Buus S (1982) Is the detection of a temporal gap frequency dependent?. J Acoust Soc Amer 71 (Suppl. 1) S48

Florentine M, Buus S (1983) Temporal resolution as a function of level and frequency. Proceedings 11th International Congress of Acoustics 3:103-106

Florentine M, Buus S (1984) Temporal gap detection in sensorineural and simulated hearing impairment. J Speech & Hearing Research 27:449-455

Florentine M, Buus S, Scharf B, Zwicker E (1980) Frequency selectivity in normal-hearing and hearing-impaired observers. J Speech & Hearing Research 23:646-669

Giraudi-Perry DM, Salvi RJ, Henderson D (1982) Gap detection in hearing-impaired chinchillas. J Acoust Soc Amer 72:1387-1393

Green DM (1973) Temporal acuity as a function of frequency. J Acoust Soc Amer 54:373-379

Green DM, Shelton BR, Picardi MC, Hafter ER (1981) Psychophysical tuning curves independent of signal level. J Acoust Soc Amer 69:1758-1762

Hellman R, Scharf B (1984) Acoustic reflex and loudness. In The Acoustic Reflex: Basic Principles and Clinical Applications (ed Silam S). Academic Press, New York

Irwin RJ, Hinchcliff LK, Kemp S (1981) Temporal acuity in normal and hearing-impaired listeners. Audiology 20:234-243

Irwin RJ, Kemp S (1976) Temporal summation and decay in hearing. J Acoust Soc Amer 59:920-925

Irwin RJ, Purdy SC (1982) The minimum detectable duration of auditory signals for normal and hearing-impaired listeners. J Acoust Soc Amer 71:967-974

Moore BCJ, Glasberg BR (1983) Suggested formulae for calculating auditory-filter bandwidths and excitation patterns. J Acoust Soc Am 74:750-753

Munson WA (1947) The growth of auditory sensation. J Acoust Soc Amer 19:584-591

Penner MJ (1975) Persistence and integration: Two consequences of a sliding integrator. Perception & Psychophysics 18:114-120. See also Penner MJ (1976) Erratum. Perception & Psychophysics 19:469-470

Penner MJ (1977) Detection of temporal gaps in noise as a measure of the decay of auditory sensation. J Acoust Soc Am 61:552-557

Plomp R (1964) Rate of decay of auditory sensation. J Acoust Soc Am 36:277-282

Scharf B (1970) Critical Bands. In Foundations of Modern Auditory Theory Vol 1 (ed. Tobias JV) pp 157-202. Academic Press, New York

Scharf B, Buus S (in press) Audition I. In Handbook of Perception and Human Performance (ed. Boff KR). John Wiley & Sons, New York

Scharf B, Hellman RP (1966) A model of loudness summation applied to impaired ears. J Acoust Soc Am 40:71-88

Shailer MJ, Moore BCJ (1983) Gap detection as a function of frequency, bandwidth, and level. J Acoust Soc Am 74:467-473

Steinberg JC, Gardner MB (1937) The dependence of hearing impairment on sound intensity. J Acoust Soc Am 9:11-23

Tyler RS, Summerfield Q, Wood EJ, Fernandes MA (1982) Psychoacoustic and phonetic temporal processing in normal and hearing-impaired listeners. J Acoust Soc Am 72:740-752

Zwicker E (1961) Subdivision of the audible frequency range into critical bands (Frequenzgruppen). J Acoust Soc Am 33:248

Zwicker E, Feldtkeller R (1967) Das Ohr als Nachrichtenempfänger. S Hirzel Verlag, Stuttgart

Zwicker E, Wright HN (1963) Temporal summation for tones in narrow-band noise. J Acoust Soc Am 35:691-669

Zwislocki JJ (1969) Temporal summation of loudness: An analysis. J Acoust Soc Am 46:431-441

Range Determination by Measuring Time Delays in Echolocating Bats

Hans-Ulrich Schnitzler, Dieter Menne and Heidi Hackbarth

Lehrstuhl Zoophysiologie, Institut für Biologie III,
Auf der Morgenstelle 28, D-7400 Tübingen, FRGermany

INTRODUCTION

Like the technical systems of radar and sonar, the echolocation systems of bats consist of a transmitter which produces and radiates a
particular type of signal and a receiver which evaluates the
returning echoes. Using echolocation, bats are able to localize a
target by measuring its distance and angular position. Additionally
they can get information on radial velocity and on target properties like size, shape and surface texture. The localization and the
characterization of the target is hampered by noise, clutter,
signals from other bats and signal changes due to atmospheric
influences.

Bats with different life habits have to solve different echolocation tasks. So bats hunting for insects in the open do not have the
problem to separate the insect echoes from background clutter as
have bats which hunt for insects in the vicinity of trees or for
insects sitting on surfaces. Bats that feed on fruit, pollen,
nectar, small vertebrates, surface fishes or on blood have other
problems to overcome.

The echolocation systems of the more than 800 different species of
echolocating microchiroptera bats have been adapted during evolution for the collection of relevant species specific information
and for the suppression of unwanted interference. This adaptation
led to different echolocation systems in different species. Such
differences can be found when comparing the echolocation signals of
different species of bats.

Structure and patterning of echolocation signals

The echolocation signals of bats are characterized by differences
in frequency structure, duration and sound pressure level (SPL).
Usually they consist either of short, broadband, frequency-modulated (fm) components alone, or of combinations of those with short
to long, narrowband, constant-frequency (cf) components. In a few
cases pure cf-signals have been observed. Further variations result
from different harmonic content. The fm-signals are mostly rather
short with durations below 5-10 ms, whereas cf-components from a
few milliseconds up to more than 100 ms have been observed.

Within one species the sound structure can be varied in different
orientation situations. Fig. 1A gives an example of a typical sound
sequence of a bat with rather long cf-fm signals emitted while

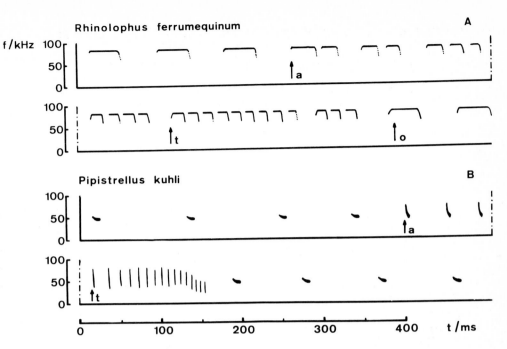

f/kHz

Fig. 1. Patterning of echolocation signals. A: Sound pattern of a _Rhinolophus ferrumequinum_ while passing an obstacle of vertically stretched wires. B: Sound pattern of a _Pipistrellus kuhli_ when hunting for an insect. The approach phase begins at "a" and the terminal phase begins at "t".

passing an obstacle of vertically stretched wires. Fig. 1B demonstrates a sequence of a typical fm bat with shallow and steep fm sweeps emitted when pursuing an insect.

The sound patterns produced in comparable situations by different species of bats are rather similar. For their characterization we will use the three behavioral categories of Griffin et al. (1960) to describe the sound patterns of hunting vespertilionid bats. A bat in free flight and in the search phase does not react to a target by an increase of its pulse emission rate. In this situation bats produce mostly a single sound of maximal duration per respiratory cycle and wing beat. Repetition rates of 4-12 pulses per second have been observed. The bat enters the approach phase when it reacts to a target by producing more and shorter sounds. Often the sounds are arranged in groups still showing correlation between respiratory cycle, wing beat and sound emission. The terminal phase is emitted when the bat is close to a target. It is characterized by a long group of very short pulses with maximal repetition rate. After a bat has caught an insect or passed an obstacle it switches back to the free flight pattern and the search phase.

Detailed reviews describing the pulse structure and the emission pattern of different species of bats have been published by Griffin (1958), Airapetianz and Konstantinov (1974), Schnitzler (1973,

1978), Novick (1977), Simmons, Fenton and O'Farrell (1979), Pye (1980), Schnitzler and Henson (1980), and Neuweiler (1983).

THE TIME DELAY HYPOTHESIS

In this paper we want to deal with the problem how bats use their orientation sounds to measure target range. For the localization of a target bats have to determine the range and the angular position of the target. Hartridge (1945a,b) was the first who assumed that bats determine the range of targets by measuring the time between sound emission and echo reception. Griffin (1944) had anticipated this principle by comparing echolocation in bats with the technical systems of radar and sonar which also use the time difference between outgoing and returning signal for range determination. This hypothesis is now widely accepted for all bats. For cf-fm bats it was suggested that they use only the fm-component of their sounds for target ranging (Schnitzler 1968, 1970, 1973, Simmons 1973).

Another hypothesis was proposed by Pye (1960) and Kay (1961). They suggested that in bats with fm-signals range information is encoded in beat notes arising by the outgoing sound and the overlapping echo. This beat frequency would decrease linearly with decreasing distance.

In the fm bat _Myotis lucifugus_, however, it was demonstrated that during the approach and terminal phase of an insect pursuit the sound duration decreases with decreasing target range so that no overlap occurs between the outgoing sound and the returning echo (Cahlander et al. 1964). Since in other fm bats and also in cf-fm bats no overlap of the fm-components occurs the beat note hypothesis can also be rejected for all other species (Schnitzler and Henson, 1980).

Range accuracy versus range resolution

When judging the bats' ability to determine range we have to distinguish between accuracy of range determination and range resolution.

Accuracy of range determination describes the ability with which a bat can measure the range of a single target or - in other words - the ability with which a bat brain can determine the time delay between outgoing sound and returning echo.

Range resolution describes the ability of a bat to decide whether the returning signal contains only one echo wavefront from a single reflecting surface or two or more such wavefronts from closely spaced surfaces.

For the discussion of the problem of how accurately bats determine distances we assume that bats use some kind of stop watch which is started by the output of the auditory receiver when processing the outgoing sound and stopped by the corresponding output produced by the returning echo. In our example we will assume that the maxima of the outputs of a hypothetical auditory receiver can be used as estimates for the moment of pulse emission (t_p) and the moment of echo reception (t_e) (Fig. 2A).

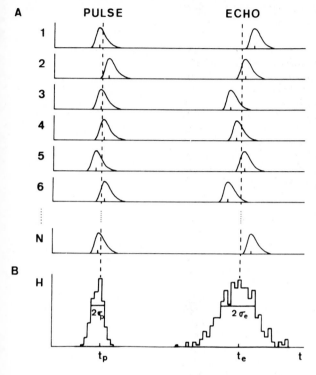

A

PULSE   ECHO

1

2

3

4

5

6

N

B

H

$2\sigma_p$ $2\sigma_e$

t_p t_e t

Fig. 2. Accuracy of time determination with a stop watch. A: Amplitude course of hypothetical receiver output at repeated presentations of pulse-echo pairs. B: Normalized histograms H when the amplitude peaks were used as time estimates for t_p and t_e.

When measuring time delays with a stop watch there will be a time error at the start and at the stop of the watch. The time estimates will be grouped around the exact time when the sound is emitted (t_p) or the echo returns (t_e). The magnitude of the error depends on the receiver characteristics of the auditory system, the signal type and the internal and external noise.

The accuracy of the system cannot be determined by a single time measurement alone. By averaging over repeated measurements we get a maximal frequency of estimates near t_p and t_e. The standard deviation of these distributions (σ_p, σ_e) can be used as a measure of the accuracy of time determination (Fig. 2B).

In this paper we want to discuss the problem how precisely bats can determine the time delay between t_p and t_e.

Range accuracy in insect catching bats

Films from insect catching bats show that they know rather precisely where their prey is located in space. Webster (1963) estimated that bats are able to pinpoint their prey in space within about one cubic centimeter. This accuracy is high enough for successful interceptions. The use of the tail membrane and the wings for capturing flying insects increases the area in which a bat can seize its prey.

Trappe (1982) trained Greater Horseshoe Bats (<u>Rhinolophus ferrumequinum</u>) to catch flour powdered mealworms which were thrown towards the bats and recorded the flour marks on the wing membrane. He found marks within an area at the base of the 3rd to the 5th finger with a diameter of about 2-3 cm. This suggests a localization accuracy of at least the same value.

Range difference threshold

More precise data concerning the accuracy with which bats are able to determine range were collected by Simmons (1971, 1973) and Airapetianz and Konstantinov (1974). In a two-alternative forced-choice experiment bats were trained to decide which of two real or simulated targets offered in different directions and at different absolute ranges was closer (Fig. 3A). The discrimination performance was measured as the percentage of correct responses and the discrimination threshold was set at the range (time) difference where the bats reached a score of 75%.

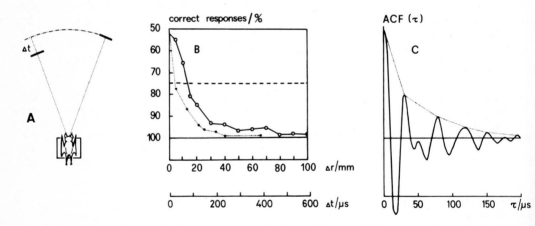

Fig. 3. Determination of range discrimation threshold in Eptesicus fuscus. A: Experimental set-up with real targets differing in range. B: Comparison of the measured average performance of 8 bats (solid line) with the envelope curve of the auto-correlation function (acf) (dotted line). C: Acf of a bat sound (solid line) with the envelope curve (dotted line). Adapted from Simmons (1973).

In these range difference experiments the bats scanned from the closer to the more distant target emitting several sounds per scan. They therefore had the opportunity to make several range measurements before making a decision.

Species	Δr/mm	Δt/μs	r/cm	Author
Eptesicus fuscus	13	75	60	Simmons 1971
Eptesicus fuscus	12	70	30	Simmons 1973
Eptesicus fuscus	14	81	240	Simmons 1973
Eptesicus fuscus	–	60	30	Simmons 1973
Phyllostomus hast.	12	70	60	Simmons 1971
Phyllostomus hast.	12	70	120	Simmons 1973
Pteronotus suapur.	15	87	30	Simmons 1973
Pteronotus suapur.	17	98	60	Simmons 1973
Rhinolophus ferr.	25	145	30	Simmons 1973
Rhinolophus ferr.	41	240	100	Airapetianz,
Myotis oxygnathus	8	46	100	Konstantinov 1974
Myotis oxygnathus	23	133	100	

Table 1. Range difference thresholds measured by different authors.

In different species of bats range difference thresholds between 8 and 25 mm have been measured. This corresponds to time difference thresholds between 46 and 145 μs (Table 1). Fig. 3B shows the average performance curve of 8 <u>Eptesicus fuscus</u> which reached an average range difference threshold of 12 mm (70 μs).

The range discrimination threshold did not change when the simulated targets were presented successively instead of simultaneously (Simmons and Lavender 1976). This indicates that absolute ranges (time delays) are measured and compared and that the decision is not made by comparison of the two echoes from the simultaneously presented targets.

An astonishing result of the range difference experiments with Eptesicus is that the threshold did not change at different absolute ranges of the targets (Table 1). This would mean that range difference discrimination is independent of echo SPL over a large range since echoes returning from a target at 240 cm should be at least 36 dB less intense than echoes from the same target at 30 cm.

Range jitter threshold

In another approach Simmons (1979) improved his method to measure the accuracy of range determination in <u>Eptesicus</u> <u>fuscus</u>. In these experiments the bats had to discriminate between two phantom targets which were created by playing back the bats' orientation sounds through two loudspeakers. One target was fixed in space at a distance of 50 cm (3 ms time delay). The other target was moved back and forth on alternate cries by adding a fixed time delay so that the phantom target was jittering by a time difference of Δt. A comparator determined which microphone received the stronger cry and enabled only this channel to return echoes (Fig. 4A). The bats' discrimination performance was measured for a jitter range of Δt from 80 down to 0 μs (Fig. 4B). The bats scanned back and forth between the two phantom targets and emitted between 1-20 pulses

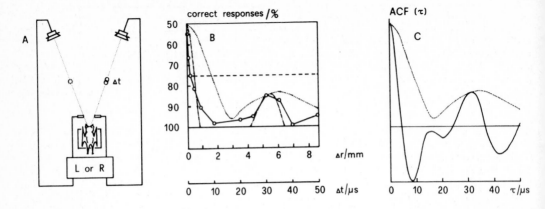

Fig. 4. Determination of range jitter threshold in *Eptesicus fuscus*.
A: Experimental set-up with a fixed and a jittering phantom target. B:
Comparison of the measured performance curve (solid line) of one bat
with the full-wave (dotted line) and half-wave (dashed line) rectified
auto-correlation function (acf) C: acf of a bat sound (solid line)
with the full-wave rectified envelope (dotted line). Adapted from
Simmons (1979).

before making a decision. They emitted the typical orientation
sounds of <u>Eptesicus</u> <u>fuscus</u> which were about 1-1.5 ms long and
contained 2 to 3 harmonics of a fundamental sweep from 55-23 kHz.

The advantage of this experiment is that the bats did not have to
compare distances of targets positioned at different angles. Therefore Simmons concluded that the range jitter threshold is less
affected by head movements which would change the target range
(Simmons and Stein 1980).

The bats discriminated the jittering echoes from the non-jittering
ones with astonishingly high accuracy. The 75% criterion for time
jitter (Δt) was about 1 μs (Simmons et al. 1983) at an absolute
time delay of 3 ms. This corresponds to a range jitter threshold
(Δr) of 0.17 mm at an absolute range of 50 cm. That means that the
range jitter threshold is about 70 times better than the range
difference threshold where the bats had to decide which of two
simultaneously presented targets was closer. Another result showed
that the bats' performance was poorer at a Δt of about 30 μs than
at shorter and longer Δt's (Fig. 4B).

How can the bats reach such a high accuracy? From Fig. 2 which
demonstrates the stop watch problem we know that the output of the
auditory receiver determines the accuracy with which bats are able
to measure time delays. This output is determined by the receiver
characteristics, the type of signal, and also by the external and
internal noise. There have been many speculations as to what type
of receiver would be best suited for a high accuracy in time
determination for a given signal type.

THE CONCEPT OF AN "OPTIMAL RECEIVER"

Strother (1961, 1967) pointed out that the signals of fm bats are rather similar to the signals used in chirp or pulse compression radar systems. He therefore assumed that echo processing in bats might be comparable to that in the matched filter receiver of a pulse compression radar system. In such receivers the returning echo is crosscorrelated with the transmitted signal by passing it through the receiver filter which is matched to the signal.

Many aspects of such optimal filtering in echolocating bats have been discussed by Cahlander (1963, 1964, 1967), van Bergejk (1964), McCue (1966, 1969), Altes and Titlebaum (1970), Altes (1973, 1975, 1976, 1980, 1981, 1984), Glaser (1971a,b, 1974) and Beuter (1976, 1977). From all these theoretical studies the only result indicating that bats might use optimal filters was that the echolocation signals of <u>Myotis</u> <u>lucifugus</u> would be optimally Doppler tolerant if they were processed in a matched filter receiver.

A matched filter receiver or optimal receiver produces the cross-correlation functions (ccf) between a replica of the emitted signal, which is stored in the matched filter, and the signals picked up by the receiving antennae. It concentrates the signal energy within a small time interval at the filter output. The pulse compression is evident if one compares a typical fm-sound of Eptesicus with its autocorrelation function, acf (Figs. 5A and C). In this case the 2 ms long pulse is compressed into an acf with a central peak which is only a few μs wide.

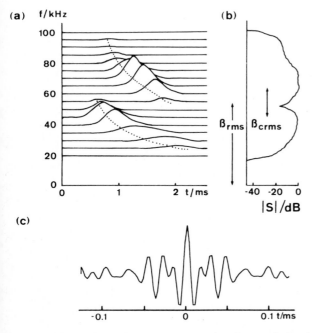

Fig. 5. Echolocation sound of an Eptesicus fuscus. A: Frequency structure of the sound. B: Energy spectrum with the rms-bandwidth (B_{rms}) and the centralized rms-bandwidth (B_{crms}).| S| Spectral amplitude. C: Auto-correlation function (acf) of the sound.

The time compressed ccf which is used to estimate the echo arrival time allows an optimal time measurement. Therefore the ccf plays an important role in all discussions of whether bats use an optimal receiver for time estimation.

In all publications it was assumed that echo and emitted pulse are rather similar so that the acf of the emitted pulse could be used as a good estimate for the unknown ccf between emitted pulse and returning echo. The acfs have been determined for the echolocation signals of many species of bats and have been discussed under the assumption that bats have some kind of neural representation of this autocorrelation function which they can use for time measurements (e.g. Simmons and Stein 1980).

When discussing the performance of an optimal receiver it is important to distinguish whether the receiver uses the ccf with its full fine structure or whether it uses only the envelope of the ccf for time estimation. A receiver which uses the full fine structure of the ccf is called a fully coherent optimal receiver. As the fine structure depends on the phase this receiver type conserves the phase information of the echoes. A receiver which uses the envelope of the ccf could be called a semicoherent optimal receiver (Menne and Hackbarth, in press). In this receiver the fine structure of the acf is eliminated by the envelope formation.

The attempt to prove an "optimal receiver" in bats

Simmons (1973, 1979) used his behavioral data of the range difference and the range jitter experiments to make conclusions on the type of information processing used by the bats' receiver when making time estimations.

In the range difference experiment Simmons (1973) argued that the bats' hearing system is matched to the waveform of their signals. Therefore the ccf between echo and emitted signal should be a measure of the accuracy with which bats are able to determine the arrival time of echoes. At this time he further assumed that with the high frequencies used by bats, phase information is not available in the auditory system so that the bats could only use the envelope of the ccf for time estimation. He therefore suggested that the "envelope curve" which connects the peaks of the acf (used as estimate for the ccf) predicts the performance of the bats in the range difference experiment (Fig. 3C). The comparison of the average performance curve of 8 Eptesicus fuscus with the envelope curve shows a good correspondence (Fig. 3B). Such a correspondence was also found when comparing the envelope of the acfs of entire sounds in Phyllostomus hastatus, a bat which produces pure fm signals, and in Pteronotus suapurensis which produces short cf-fm sounds. In Rhinolophus ferrumequinum which uses long cf-fm signals only, the acf of the fm sweep alone produces a corresponding prediction curve. This close fit between the envelopes of the acfs and the performance curves led Simmons (1973) to the conclusion that echolocating bats have a neural equivalent of a matched filter receiver (or a semicoherent optimal receiver in our terminology).

Another argument for the existence of a semicoherent optimal receiver was that the accuracy of range determination depends on the "sonagram-bandwidth" of the orientation sounds (measured as distance between lowest and highest frequency in the sonagram) as this bandwidth somehow determines the shape of the crosscorrelation function. When comparing this bandwidth with the range difference threshold Simmons et al. (1975, 1983) pointed out that time difference at threshold is approximately equal to the reciprocal of the sonagram-bandwidth of the echolocation signals. This also led to the conclusion that bats behave as though they possess the biological equivalent of a crosscorrelating receiver.

When interpreting his range jitter experiment Simmons (1979) used the same line of argument as in the range difference experiment. He also compared the measured data with the acf of the emitted signal (Fig. 4C). Now he showed that the bats perform as though they perceive a half-wave rectified version of the acf and not the full wave rectified envelope (Fig. 4B). From the close fit with the fine structure of the acf and especially from the reduced performance at the first side peak of the acf at about 30 µs he concluded that the bats perceive small time intervals well enough to sense phase information in the echoes. He assumed that the acf with its full fine structure is the most appropriate representation of the sonar signal in the bats' brain and that the acf can be used to discuss the contribution of the signals to range perception. That means that bats would use a fully coherent receiver.

As explanation for the different results in the range discrimination and the range jitter experiments Simmons and Stein (1980) suggested that in the range discrimination experiment the bats' trial-by-trial head movements obscured the acf itself, and only the envelope was available to the bat. Thus by scanning, the bat could only get range information corresponding to the performance of a semicoherent receiver whereas in the jitter experiment bats behave as if they have a fully coherent receiver which makes use of the fine structure of the acf. Simmons suggested that bats with broad band fm-signals use temporal information not only for ranging but also for range resolution and angular localization (Simmons 1979, Simmons et al. 1983).

There is no proof for an "optimal receiver"

The argument that optimal receivers can utilize the ccf between signal and echo to determine the accuracy of time (range) is correct. However, the shape of the acf cannot be used for predicting the performance of bats in ranging experiments as assumed by Simmons. If a bat were to use the acf or the envelope of the acf to estimate the moment of pulse emission and echo reception, the accuracy of time determination would be a monotonic function of the signal-to-noise-ratio, SNR (Schnitzler and Henson 1980, Schnitzler 1984). For a good SNR, the standard deviations of the distributions of time estimates (σ_p and σ_e) are far smaller than the width of the acf.

It is a possibility, however, to predict the accuracy of time determination for coherent and semicoherent receivers if the bandwidth of the signal (β), the signal energy (E) and the noise power

per unit bandwidth (N_o) are known. For this prediction different
bandwidths must be used for the coherent and the semicoherent
receiver (Hackbarth 1984, Menne and Hackbarth in press). The rms-
bandwidth of the coherent receiver (β_{rms}) is a weighted distance of
frequencies in the power spectrum relative to 0 Hz (Burdic 1968).
The centralized rms-bandwidth of the semicoherent receiver (β_{crms})
is a measure of the frequency extent of a power spectrum (Altes
1971). For both receivers the accuracy of time determination σ can
be calculated for a reasonable signal-to-noise ratio (Burdic 1968):

$$\sigma = \frac{1}{2\pi\beta_x \sqrt{\frac{2E}{N_o}}}$$

For the calculation of σ_{rms} of a coherent receiver β_X has to be
replaced by β_{rms} and for σ_{crms} of a semicoherent receiver by β_{crms}.
An echolocation signal of Eptesicus that produces an acf similar to
the one shown by Simmons (1979) has a β_{rms} of 52.7 kHz and a β_{crms}
of 15.7 kHz (Fig. 5). This sound has a duration of about 1.5 ms and
consists of 2 harmonics with the first harmonic sweeping from 55 to
23 kHz. We used the β_{rms} and β_{crms} of this signal to calculate σ_{rms}
and σ_{crms} for different signal-to-noise ratios (Table 2). The
results show that in both receiver types σ depends strongly on the
SNR. A change in SNR of 20 dB corresponds to a ten fold change in
σ. Table 2 shows that a coherent receiver is about 3 times more
accurate than a semicoherent receiver.

Eptesicus fuscus	coherent receiver β_{rms} = 52.7 kHz	semicoherent receiver β_{crms} = 15.7 kHz
Signal to noise ratio $\sqrt{2E/N_o}$	σ_{rms} / µs	σ_{crms} / µs
10 (20 dB)	0.302	1.014
100 (40 dB)	0.030	0.101
1000 (60 dB)	0.003	0.010

Signal to noise ratio	range jitter Δt / µs	range difference Δt / µs
at least 40-60 dB	1	60 - 80

Table 2. Comparison of the time measurement error (σ) of a semicoherent and a coherent receiver at different SNRs with the range jitter and range difference threshold as measured by Simmons (1971, 1973, 1979).

How do these predictions compare with the behavioral data of
Simmons? The comparison is only possible if we know the SNR of the
returning echoes in the range difference and the range jitter
experiments. In Figs. 11 and 12 of the range difference paper
Simmons (1973) shows oscillograms of the bats outgoing sonar
signals and of the returning echoes. In order to make a comparison
possible we simulated different echo SNRs with an Eptesicus pulse.
At a SNR of 40-60 dB we got a similar oscilloscope trace. Therefore
we assume that the behavioral experiments were made at this SNR. A
comparison of the calculated σ_{rms} and σ_{crms} for the determination
of t_p and t_e with the measured range difference and range jitter
threshold shows that the predictions are far smaller than the
measured behavioral thresholds for a SNR of 40-60 dB.

However, a direct comparison of the predicted temporal accuracy
with behavioral thresholds in the range difference and in the range
jitter experiments is not possible. For instance in the range
difference experiment a bat has to make at least four time measurements in order to make a decision (at each distance a measurement
when starting the stop watch and another when stopping it). However, an estimation of the total error shows that the predicted
accuracy for the range jitter and the range difference experiments
is still far better than the behavioral performance measured by
Simmons (see Table 2). This means that the results of the
behavioral experiments do not prove the use of a semicoherent or
coherent optimal filter receiver by the bats, and that there is no
proof for a neuronal equivalent of a matched filter or crosscorrelation receiver in bats.

Menne and Hackbarth (in press) and Hackbarth (1984) have simulated
with a laboratory computer the decision procedure necessary to
compare the distances in the range jitter and range difference
experiments in order to get prediction curves which are comparable
to the behavioral results. Their simulations also showed that the
behavioral data do not prove the existence of an optimal receiver
in bats.

Another line of argument for a semicoherent optimal receiver was
that the range difference thresholds of different species of bats
depend on the sonagram-bandwidths of the bats' signals and are
equal to the reciprocal of these bandwidths (Simmons et al. 1975,
Simmons et al. 1983). The formula for predicting temporal accuracy
shows that the sonagram-bandwidth can only be used as an estimate
of σ if:
1. the sonagram-bandwidth is an estimate for either β_{rms} or β_{crms}.
With sufficient accuracy the sonagram-bandwidth gives an estimate
for β_{crms} which is about 1/3 to 1/4 of the sonagram-bandwidth. This
bandwidth can only be used to predict the performance of a semicoherent receiver.
2. the SNRs of the compared time measurements are similar;
3. the SNRs are good, since the formula makes predictions only at
reasonable SNRs (Menne and Hackbarth, in press).

The above prerequisites are not fullfilled, since:
1. the sonagram-bandwidth used was about 3-4 times larger than

β_{crms}. For Eptesicus an incorrect sonagram-bandwidth of 25 kHz was used by Simmons et al. (1975), but a bandwidth of 70-80 kHz was described later for this species (Simmons 1979);

2. the SNR of the range difference experiments in different species was surely different according to different SPLs of the signals used and according to different absolute ranges (Schnitzler and Henson 1980);

3. the reciprocal relation between β_{crms} and σ would mean a $\sqrt{2E/N_o}$ of 1 or a SNR of 0 dB. At this SNR the formula cannot be used. It is also evident that the range difference thresholds must have been determined at far better SNRs. Consequently the bandwidth argument cannot be used to prove that bats have an optimal semicoherent receiver. Even the tendency that bats with broadband signals have a better range difference threshold than bats with signals of smaller bandwidth has not been proven.

When comparing the range jitter performance curve with the acf Simmons (1979) found a good fit to the half-wave rectified version of the acf. From this close fit, and especially from the reduced performance at about 30 μs which reproduces the ambiguity of the first sidepeak of the acf, he concluded that bats sense phase information in sonar signals. They behave as if they have an fully coherent optimal receiver. The question whether this experimental approach allows such farreaching conclusions is again a matter of conjecture (Schnitzler 1984, Hackbarth 1984, Menne and Hackbarth, in press). A possible explanation for the extremely small jitter threshold and for the peak in the performance curve at about 30 μs could be that the bats perceived a non-jittering scatter echo from the surroundings together with the jittering echoes. They may have used interference pattern between the scattered and the direct echoes originating at different Δt's. In this case bats could use spectral cues and would not need phase information for the jitter discrimination.

Another argument against the use of a semicoherent receiver in the range difference experiment and a fully coherent receiver in the range jitter experiment is that the two receiver types differ in performance by a factor of 3 (see Table 2). The behavioral data show a 60-80 times threshold change in the range difference experiment relative to the range jitter experiment.

All these arguments lead us to the conclusion that there is absolutely no proof for optimal information processing in bats. We see no argument so far which allows to use the acfs of echolocation signals to make any statements about the accuracy in range determination in bats. We think that the correlations found are more or less accidental. Therefore we suggest the use of acfs of echolocation signals be discontinued or used critically until there is a more convincing argumentation on the relationship between the behavioral performance of bats in ranging experiments and the acfs of their sounds.

RANGE RESOLUTION

Range resolution describes the ability of bats to decide whether a
returning echo contains only one wavefront from a single reflecting
surface or several wavefronts from closely spaced surfaces. This
ability may be important when bats identify targets according to
their surface structure. It also plays a role when bats have to
separate weak target echoes from nearby interfering clutter echoes.
The range resolution performance of bats could be measured in a
two-alternative forced-choice experiment where bats have to discriminate a one-wavefront target (e.g. a flat plate) from a two-
wavefront target with variable time delays between the two wavefronts (e.g. a flat plate with holes of different depths). Such an
experiment has not yet been conducted.

In a related experimental approach (Simmons et al. 1974) bats were
trained to discrimate a two-wavefront control target (plate with
holes of 8 mm depth) with a fixed delay of 46 μs from two-wavefront
test targets with variable delays. In this situation <u>Eptesicus</u>
<u>fuscus</u> was able to discriminate the test target when the holes were
0.6-0.9 mm shallower than the 8 mm deep holes of the control plate
(Simmons et al. 1974). This corresponds to a difference of the
two-wavefront delays of 3.5-5.2 μs. In <u>Myotis</u> <u>myotis</u> a range
difference of 1 mm (6 μs) was necessary at 8 mm deep holes and of
0.8 mm (4.6 μs) at 4 mm deep holes of the control plate (Habersetzer et al. 1983). In <u>Megaderma</u> <u>lyra</u> a depth difference of 1.4 mm
(8 μs) was necessary for discrimination at 8 mm deep holes of the
control plate (Vogler and Leiner, unpublished data as cited in
Habersetzer and Vogler 1983).

What are the cues used by the bats in this type of discrimination
experiment? Simmons et al. (1974) and Beuter (1980) demonstrated
that the two overlapping echoes returning from a two-wavefront
target produced an interference pattern which is characterized by a

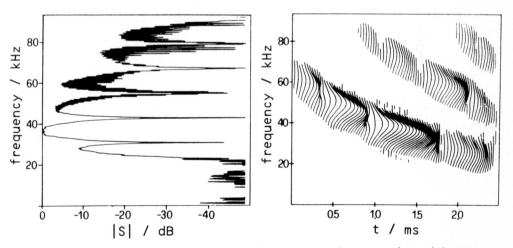

*Fig. 6. Simulated echo from a simple structured target formed by two
point targets separated in depth by 14 mm. Left: Spectral amplitude S.
Right: Frequency structure of the simulated echo. Adapted from Beuter
(1980).*

pattern of periodic minima. Under some simplifying assumptions the frequency separation Δf between the minima is related to the time delay Δt between the two wavefronts by Δf = 1/ Δt. Fig. 6 shows for Eptesicus fuscus the power spectrum and the spectrogram of a simulated echo from two single-point targets with a depth difference of 14 mm (80 μs). At this time delay the minima in the spectrum have a separation of 12.4 kHz. Simmons et al. (1974) presented the spectra of the sound reflected by real targets used in the hole-depth discrimination experiment. These spectra are also characterized by depth-dependent patterns of periodic minima which differ, however, from the calculated spectra of simulated two-wavefront echoes from point targets. This may be due to interference and resonance phenomena produced by the geometrical structures of the hole-plate. Simmons et al. (1974), Beuter (1980) and Habersetzer and Vogler (1983) suggested that the bats use the spectral cues of two-wavefront echoes to discriminate between the targets with different hole-depths. An Eptesicus fuscus which is able to discriminate a control plate with 8 mm holes from a test plate with 7.4 mm holes should therefore be able to recognize the different minima pattern in the spectra of the two-wavefront echoes.

After the jitter experiment Simmons (1979) modified his interpretation and suggested that information about echo arrival time could have served as cues for the bat. He assumed that the extreme acuity of ranging in the range jitter experiment shows that bats with broadband echolocation sounds use temporal information for most, if not all perceptual tasks. An Eptesicus fuscus should thus be able to discriminate time delays between two overlapping echoes of 46 μs and 42.5 μs respectively. A delay difference of 3.5 μs is sufficient for the bat to discriminate between the two echo patterns.

We think that it is not allowed to use the threshold of a range accuracy experiment, where a bat has to determine the time delay between an outgoing pulse and non overlapping echos, as an indication that bats must be able to measure the time delay between two overlapping echoes with about the same accuracy. We see no reason why the bats should use only temporal cues and not the available spectral cues.

THE NEURONAL BASIS OF RANGE DETERMINATION

If we assume that the time delay between emitted sound and returning echo encodes the target range we must address the question of how this delay is represented within the bats' auditory system.

Neurophysiological studies in the inferior colliculus of Myotis lucifugus (Suga 1970) showed accurately time locked neurons which fired with one spike at a nearly constant latency to a tonal stimulus regardless of the stimulus amplitude and envelope. Such neurons are particularly specialized to encode time information. Neurons of this type have not been found on the level of the cochlear nucleus where spontaneously and nonspontaneously active neurons fired repetitively in response to a tonal stimulation.

Similar latency-constant neurons which often selectively fire only at fm-stimuli and show great temporal precision of firing, have also been reported for <u>Tadarida</u> <u>brasiliensis</u> (Pollak et al. 1977a,b), <u>Mollossus</u> <u>ater</u> and <u>Molossus</u> <u>molossus</u> (Vater and Schlegel 1979) and <u>Rhinolophus</u> <u>ferrumequinum</u> (Vater 1981). These neurons can serve as time markers.

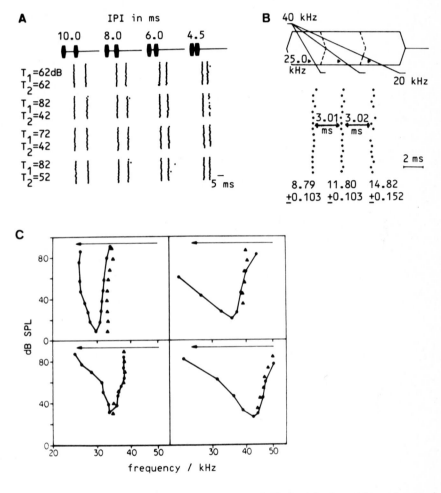

Fig. 7. Responses of collicular neurons of <u>Tadarida</u> <u>brasiliensis</u> to fm pulses or pulse-echo pairs. A: Response pattern to fm pulse- echo pairs of different sound pressure levels (T1, T2) and different interpulse intervals (IPI). Each pair of dot columns was generated by 16 repetitions of the simulated pulse-echo combination. Adapted from Pollak (1980). B: Response pattern to fm pulses of different duration. Each dot column represents spikes elicited by 16 presentations at a given duration. The exciting frequencies (EF) are indicated by arrows. Each dot column's actual mean response latency and standard deviation of mean latency is shown below that column (in ms). Adapted from Bodenhamer et al. (1979). C: Relationship of the exciting frequencies (EFs) at fm stimuli of different SPL and the excitatory areas of tuning curves of four collicular neurons of <u>Tadarida</u> <u>brasiliensis</u>. The arrows above each tuning curve show the sweep range of the fm stimuli. Adapted from Bodenhamer et al. (1981).

When simulating the natural echolocation situation by stimulating with fm pulse-echo pairs, Pollak and his coworkers (summarized in Pollak 1980) found neurons in the colliculus inferior of Tadarida brasiliensis which appear to be specialized for range coding. These neurons showed a phasic on-response of mostly one or sometimes two spikes to the simulated pulse and the echo with a very constant latency over a wide range of intensities. When testing such neurons with fm signals of variable duration it could be shown that the neuron always fired at the same exciting frequency (EF), a narrow frequency component of 0.5-2.5 kHz in the fm sweep (Figs. 7A,B). These neurons can serve as precise time markers which can encode the time delay between outgoing pulse and echo and convey the information to higher auditory centers.

When comparing EF at different signal intensities with the tuning curves of the tested neuron it could be shown that the cell normally fires as soon as the fm signal enters the exciting area (Bodenhamer and Pollak 1981). As such neurons normally have a steep slope on the high frequency side, the triggering at the tuning curve's edge is hardly influenced by the signal intensity (Fig. 7C).

The exact coding of the synchronized time-delay responses of neurons in the inferior colliculus does not mean that the brain has read this information. Feng et al. (1978) who studied the response of neurons to paired fm stimuli in the nucleus intercollicularis of Eptesicus fuscus got results which suggest how range information may be processed by higher auditory centers. They found neurons which did not respond to single tonal stimuli, but responded vigorously when fm pulse-echo pairs were presented at certain time delays. The delay tuning of these neurons was about 5-8 ms wide which the authors think to be too wide to explain the high accuracy of range determination found in behavioral experiments.

Delay sensitive neurons were also found in the auditory cortex of Myotis lucifugus, another fm bat (Sullivan 1982a,b). These neurons responded selectively to simulated fm pulse-echo pairs of a loud pulse and a weak echo with delay dependent facilitation (Fig. 8). They responded only if the time separating the pulse and echo was appropriate. The delay response function (percentage of maximal response at different delays) of such neurons is narrowly tuned to the best delay of the unit (Fig. 8). Since time delay encodes range it was suggested that these delay sensitive neurons play an important role in range determination in bats. The paper gives no information how the neurons are arranged in the bats' auditory cortex.

The most detailed information on range sensitive neurons is available for the cf-fm bat Pteronotus parnellii (Suga and O'Neill 1979; O'Neill and Suga 1979, 1982). In the auditory cortex of this cf-fm bat several specialized areas for the processing of different types of biosonar information have been found (latest review by Suga 1984). The orientation signals of the mustache bat consist of four harmonics each with a cf and a fm component. When simulating ranging, pulse-echo pair combinations of the different harmonics have been used as stimuli. Time-delay sensitive neurons were found in a 3 mm area dorsorostral to the tonotopically organized auditory cortex field (FM-FM area). The neurons there responded poorly to

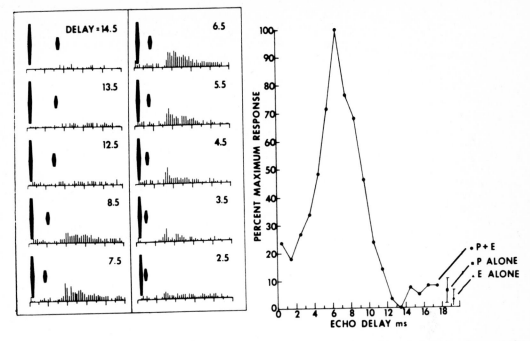

Fig. 8. Response pattern (left) and response function (right) of a delay tuned neuron from the auditory cortex of the bat Myotis lucifugus which was stimulated with fm pulse-echo pairs of different interpulse intervals. On the left are shown the pulse-echo pairs with the PST histograms of the responses. On the right is shown the spike count function for these responses. At the lower right-hand corner the average pulse alone and echo alone responses are indicated as a percentage of the maximum pair response. Adapted from Sullivan (1982a).

pure tones and fm sweeps alone but responded vigorously to pulse-echo pairs.

By systematically eliminating the various cf and fm components in each stimulus pair it was found that the most effective components for delay specific facilitation were the first harmonic of the fm sweep in the stimulus (FM_1) in combination with the second, third or fourth harmonic of the fm sweep in the simulated echo (FM_2, FM_3, FM_4).

By changing the sound pressure level of the simulated echo it was possible to determine the delay-tuning curve of the delay sensitive neurons. Best delay (BD) was defined as the delay where the neuron was most sensitive. In some of the neurons the shape of the delay-tuning curves and the best delay changed as the repetition rate and the stimulus duration were varied. These neurons which were called tracking neurons will not be discussed here. In the more common delay-tuned neurons the best delay was less variable and the tuning was sharper. Such neurons respond only when the target is at a favorable distance. Therefore they were also called range-sensitive neurons. The delay-tuning curves of most range-sensitive neurons are closed at higher echo sound pressure levels. They are not only tuned to a specific delay but also to a particular amplitude relationship between pulse and echo (Fig. 9A).

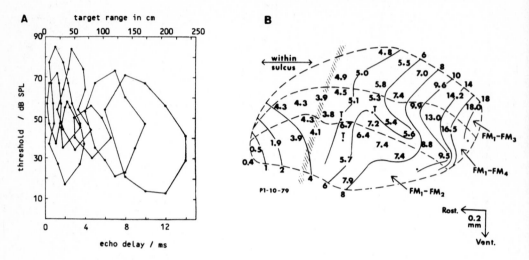

Fig. 9. Representation of range in the auditory cortex of Pteronotus parnellii. A: Delay tuning curves of seven neurons as measured with simulated pulse-echo pairs (FM_1-FM_2). At each neuron the SPL of the simulated pulse was kept constant whereas the SPL of the simulated echo was varied in order to determine the facilitation threshold. B: Distribution of best delays (BD) in the tangential plane of the FM-FM area of the left auditory cortex of one bat. The neurons are arranged in clusters which contained exclusively FM_1-FM_2, FM_1-FM_3 or FM_1-FM_4 neurons. Isocontour BDs show the increase of BD in rostrocaudal direction. Adapted from O'Neill et al. (1982).

There are three separate longitudinal clusters containing either FM_1-FM_2, FM_1-FM_3 or FM_1-FM_4 neurons. Within these clusters the delay-tuned neurons are systematically arranged with increasing best delay along the cortical surface in the rostrocaudal direction (Fig. 9B). The BDs cover a range from 0.4-18 ms which corresponds to a target range from 7 to 310 cm. This is about the range in which bats preferably react to targets (Schnitzler and Henson 1980). The delays from 3-8 ms (50-140 cm) are especially well represented.

The data obtained from Pteronotus are very important for the understanding of the neural evaluation of time delays. They show that specialized neurons tuned to particular time delays or ranges are activated along a neuronal axis corresponding to target range. It is not known, however, how this information is interpreted or used to activate adequate behavioral programs.

All neurophysiological data presented so far show that the precision of time delay coding depends somehow on the relation of the SPLs of pulse and echo as heard by the bat. What is this relation in natural echolocation situations?

ATTENUATION MECHANISMS FOR THE EMITTED SOUNDS AND FOR ECHOES WITH SHORT DELAY TIMES

Echolocating bats receive rather faint echoes shortly after the emission of sounds with high sound pressure levels. This led to the question whether the loud emitted pulses overload the auditory system and reduce the bats' ability to detect and evaluate less intense echoes.

The investigation of this problem showed that bats hear their own sounds with a reduced intensity according to the following mechanism: The SPL of the emitted sounds arriving at the eardrum is reduced due to the directionality of pulse emission and of hearing (reviewed in Schnitzler and Henson 1980). The contraction of the middle ear muscles during sound emission produces an additional reduction (Henson 1970, Pollak and Henson 1973, Suga and Jen 1975). A neural mechanism for the attenuation of self stimulation by the emitted sounds has also been found (Suga and Schlegel 1972, 1973, Suga and Shimozawa 1974). All these mechanisms together prevent an overload of the auditory system by the loud emitted sounds. The difference between the SPL with which the bat hears its own sounds and the SPL of the echoes in not so large. This may help to get precise time measurements.

The contractions of muscles in the middle ear are triggered by vocalizations, and last up to 10 ms after the onset of sound emission (Henson 1970, Suga and Jen 1975). This means that echoes returning with shorter time delays are also attenuated in the middle ear. In _Tadarida brasiliensis_ the strongest attenuation (30 dB) occurred at about the onset of the sound emission. Thereafter the attenuation decreased gradually and reached 0dB at about 8 ms after the onset. In behavioral experiments with _Eptesicus fuscus_ Simmons and Kick (1984) found corresponding reductions in hearing sensitivity at echo delays shorter than 6 ms. They pointed out that the delay dependent reduction of echo SPL of 0 to 30 dB at delays from 6 to 0 ms compensates very well for the increase of echo SPL of a spherical target positioned at corresponding distances. A bat approaching a target will thus perceive about the same echo strength at time delays below about 6 ms corresponding to target ranges closer than 1 m. By this compensation the extent of possible echo SPL's is reduced.

For range determination the auditory system of bats has to evaluate pulse-echo pairs which are characterized by pulses of moderate SPL and echoes with SPL in a much smaller extent than one would assume on the first view. This restricted extent of echo SPL's may be a further help for precise time measurements.

CONCLUSIONS

The results of experiments investigating the problem of range determination in bats allow the following conclusions:
1. Bats determine range by measuring time delays between outgoing signals and returning echoes.
2. Observations in insect catching bats and the range difference experiments indicate an accuracy of range determination of about 1-2 cm which corresponds to an accuracy of time determination of about 60-120 µs.
3. The far better accuracy of the jitter experiment (0.17 mm/1 µs) is difficult to explain. Further experiments should be made to confirm these important data and to make sure that the bats did not use other cues for discrimination.
4. There is no proof that bats have some kind of a matched filter receiver or optimal receiver. Therefore the auto-correlation functions of orientation sounds give no information about the bats' ability to determine range.
5. Neurophysiological experiments indicate that at lower levels of the auditory pathway the moments of pulse emission and echo reception are marked by synchronous responses in many parallel frequency tuned neurons. In higher auditory centers the time delay information is encoded by delay tuned or range tuned neurons. At least in Pteronotus these neurons are arranged along a range axis. The accuracy of range determination may be improved by averaging over many parallel channels.
6. A precise determination of range is also favoured by mechanisms which attenuate the SPL of the outgoing sound and which reduce the SPL of echoes from nearby targets thus creating a limited extent of possible pulse-echo SPL ratios.

ACKNOWLEDGEMENTS

This study was supported by the Deutsche Forschungsgemeinschaft (Schn 138/14-15/7). We wish to thank Dr. Lee Miller for editorial suggestions and M. Johnen for technical assistance.

REFERENCES

Airapetianz ES, Konstantinov AI (1974) Echolocation in nature. Nauka, Leningrad (in Russian) English Translation: Joint Publications Research Service, Arlington
Altes RA (1971) Methods of wide-band signal design for radar and sonar systems (Ph.D. Dissertation, University of Rochester)
Altes RA (1973) Some invariance properties of the wide band ambiguity function. J Acoust Soc Am 53:1154-1160
Altes RA (1975) Mechanism for aural pulse compression in mammals. J Acoust Soc Am 57:513-515
Altes RA (1976) Sonar for generalized target description and its similarity to animal echolocation systems. J Acoust Soc Am 59:97-105

Altes RA (1980) Detection, estimation, and classification with spectrograms. J Acoust Soc Am 67:1232-1246

Altes RA (1981) Echo phase perception in bat sonar? J Acoust Soc Am 69:505-508

Altes RA (1984) Echolocation as seen from the viewpoint of radar/sonar theory. In: Varjú D, Schnitzler H-U (eds) Localization and orientation in biology and engineering. Springer, Berlin, Heidelberg, pp 235-244

Altes RA, Titlebaum EL (1970) Bat signals as optimally tolerant waveforms. J Acoust Soc Am 48:1014-1020

Bergeijk WA van (1964) Sonic pulse compression in bats and people: a comment. J Acoust Soc Am 36:594-597

Beuter KJ (1976) Systemtheoretische Untersuchungen zur Echoortung der Fledermäuse. Thesis, University of Tübingen, FRGermany

Beuter KJ (1977) Optimalempfängertheorie und Informationsverarbeitung im Echoortungssystem der Fledermäuse. In: Hauske G, Butenandt E (eds) Kybernetik 1977, München, Wien, Oldenburg, pp 106-125

Beuter KJ (1980) A new concept of echo evaluation in the auditory system of bats. In: Busnel RG, Fish J (eds) Animal sonar systems. Plenum Press, New York, pp 747-762

Bodenhamer RD, Pollak GD (1981) Time and frequency domain processing in the inferior colliculus of echolocating bats. Hearing Res 5:317-335

Bodenhamer RD, Pollak GD, Marsh DS (1979) Coding of fine frequency information by echoranging neurons in the inferior colliculus of the Mexican free-tailed bat. Brain Res 171:530-535

Burdic WS (1968) Radar signal analysis. Prentice-Hall Inc, Englewood Cliffs N J

Cahlander DA (1963) Echolocation with wide-band waveforms. Thesis deposited in the Library of MIT, Boston, Mass

Cahlander DA (1964) Echolocation with wide-band waveforms: Bat sonar signals. MIT Lincoln Lab Techn Rep No 271, AD 605322

Cahlander DA (1967) Discussion of Batteau's paper. In: Busnel R-G (ed) Animal Sonar Systems, vol. II, Laboratoire de Physiologie acoustique, Jouy-en-Josas, pp 1052-1081

Cahlander DA, McCue JJG, Webster FA (1964) The determination of distance by echolocating bats. Nature 201:544-546

Feng AS, Simmons JA, Kick SA (1978) Echo detection and target-ranging neurons in the auditory system of the bat Eptesicus fuscus. Science 202:645-648

Glaser W (1971a) Eine systemtheoretische Interpretation der Fledermausortung. Studia Biophysica 27:103-110

Glaser W (1971b) Zur Fledermausortung aus dem Gesichtspunkt der Theorie gestörter Systeme. Zool Jb Physiol 76:209-229

Glaser W (1974) The hypothesis of optimum detection in bats echolocation. J Comp Physiol 94:227-248

Griffin DR (1944) Echolocation by blind men, bats and radar. Science 100:589-590

Griffin DR (1958) Listening in the dark. Yale University Press, New Haven

Griffin DR, Webster FA, Michael CR (1960) The echolocation of flying insects by bats. Anim Behav 8:141-154

Habersetzer J, Vogler B (1983) Discrimination of surface-structured targets by the echolocating bat Myotis myotis during flight. J Comp Physiol 152:275-282

Hackbarth H (1984) Systemtheoretische Interpretation neuerer ver-haltens- und neurophysiologischer Experimente zur Echoortung der Fledermäuse. Thesis, University of Tübingen, FRGermany

Hartridge H (1945a) Acoustic control in the flight of bats. Nature 156:490-494

Hartridge H (1945b) Acoustic control in the flight of bats. Nature 156:692-693

Henson OW Jr (1970) The ear and audition. In: Wimsatt WA (ed) Biology of Bats, Vol II. Academic Press New York, pp 181-263

Kay L (1961) Perception of distance in animal echolocation. Nature 190:361

McCue JJG (1966) Aural pulse compression by bats and humans. J Acoust Soc Am 40:545-548

McCue JJG (1969) Signal processing by the bat, Myotis lucifugus. J Auditory Res 9:100-107

Menne D, Hackbarth H (in press) Accuracy of distance measurement in the bat Eptesicus fuscus. J Acoust Soc Am

Neuweiler G (1983) Echolocation and adaptivity to ecological constraints. In: Huber F, Markl H (eds) Neuroethology and behavioral physiology: Roots and growing points. Springer, Berlin, Heidelberg, pp 280-302

Novick A (1977) Acoustic orientation. In: Wimsatt WA (ed) Biology of bats, Vol. III, pp 73-287

O'Neill WE, Suga N (1979) Target range-sensitive neurons in the auditory cortex of the mustache bat. Science 203:69-73

O'Neill WE, Suga N (1982) Encoding of target range and its representation in the auditory cortex of the mustached bat. J Neuroscience 2:17-31

Pollak GD (1980) Organizational and encoding features of single neurons in the inferior colliculus of bats. In: Busnel RG, Fish JF (eds) Animal Sonar Systems, Plenum, New York, pp 549-587

Pollak GD, Henson OW Jr (1973) Specialized functional aspects of the middle ear muscles in the bat, Chilonycteris parnellii. J Comp Physiol 86:167-174

Pollak GD, Marsh D, Bodenhamer R, Souther A (1977a) Echo-detecting characteristics of neurons in the inferior colliculus of unanesthetized bats. Science 196:675-678

Pollak GD, Marsh D, Bodenhamer R, Souther A (1977b) Characteristics of phasic-on neurons in inferior colliculus of unanesthetized bats with observations relating to mechanisms of echo ranging. J Neurophysiol 40:926-942

Pye JD (1960) A theory of echolocation by bats. J Laryngol Otol 74:718-729

Pye JD (1980) Echolocation signals and echoes in air. In: Busnel RG, Fish JF (eds) Animals sonar systems. Plenum, New York, p 309

Schnitzler HU (1968) Die Ultraschall-Ortungslaute der Hufeisen-Fledermäuse (Chiroptera-Rhinolophidae) in verschiedenen Orientierungssituationen. Z vergl Physiol 57:376-408

Schnitzler HU (1970) Comparison of the echolocation behavior in Rhinolophus ferrumequinum and Chilonycteris rubiginosa. Bijdr Dierk 40:77-80

Schnitzler HU (1973) Die Echoortung der Fledermäuse und ihre hörphysiologischen Grundlagen. Fortschr Zool 21:136-189

Schnitzler HU (1978) Die Detektion von Bewegungen durch Echoortung bei Fledermäusen. Verh Dtsch Zool Ges, Gustav Fischer Verlag, Stuttgart, pp 16-33

Schnitzler HU (1984) The performance of bat sonar systems. In: Varjú D, Schnitzler HU (eds) Localization and orientation in biology end engineering. Springer, Berlin, Heidelberg, pp 211-224

Schnitzler HU, Henson OW Jr (1980) Performance of airborne animal sonar systems: I. Microchiroptera. In: Busnel RG, Fish JF (eds) Animal sonar systems, Plenum Press, New York, pp 109-181

Simmons JA (1971) Echolocation in bats: Signal processing of echoes for target range. Science 171:925-928

Simmons JA (1973) The resolution of target range by echolocating bats. J Acoust Soc Am 54:157-173

Simmons JA (1979) Perception of echo phase information in bat sonar. Science 204:1336-1338

Simmons JA, Fenton MB, O'Farrell MJ (1979) Echolocation and pursuit of prey by bats. Science 203:16-21

Simmons JA, Howell DJ, Suga N (1975) Information content of bat sonar echoes. Am Sci 63:204-215

Simmons JA, Kick SA (1984) Physiological mechanisms for spatial filtering and image enhancement in the sonar of bats. Ann Rev Physiol 46:599-614

Simmons JA, Kick SA, Lawrence BD (1983) Localization with biosonar signals in bats. In: Ewert JP, Capranica RR, Ingle DJ (eds) Advances in vertebrate neuroethology. Plenum, New York, p 247

Simmons JA, Lavender WA (1976) Representation of target range in the sonar receivers of echolocating bats. J Acoust Soc Am 60 (suppl. 1): S5

Simmons JA, Lavender WA, Lavender BA, Doroshow CA, Kiefer SW, Livingston R, Scallet AC (1974) Target structure and echo spectral discrimination by echolocating bats. Science 186:1130-1132

Simmons JA, Stein RA (1980) Acoustic imaging in bat sonar: echolocation signals and the evolution of echolocation. J Comp Physiol 135:61-84

Strother GK (1961) Note on the possible use of ultrasonic pulse compression by bats. J Acoust Soc Am 33:696-697

Strother GK (1967) Comments on aural pulse compression in bats and humans. J Acoust Soc Am 41:529

Suga N (1970) Echo-ranging neurons in the inferior colliculus of bats. Science 170:449-452

Suga N (1984) Neural mechanisms of complex-sound processing for echolocation. Neurosciences 7:20-27

Suga N, Jen PHS (1975) Peripheral control of acoustic signals in the auditory system of echolocating bats. J Exp Biol 62:277-311

Suga N, O'Neill WE (1979) Neural axis representing target range in the auditory cortex of the mustache bat. Science 206:351-353

Suga N, Schlegel P (1972) Neural attenuation of responses to emitted sounds in echolocating bats. Science 177:82-84

Suga N, Schlegel P (1973) Coding and processing in the auditory systems of FM-signal-producing bats. J Acoust Soc Am 54:174-190

Suga N, Shimozawa T (1974) Site of neural attenuation of responses to selfvocalized sounds in echolocating bats. Science 183:1211-1213

Sullivan WE (1982a) Neural representation of target distance in auditory cortex of the echolocating bat <u>Myotis</u> <u>lucifugus</u>. J Neurophysiol 48:1011-1032

Sullivan WE (1982b) Possible neural mechanisms of target distance coding in auditory system of the echolocating bat <u>Myotis</u> <u>lucifugus</u>. J Neurophysiol 48:1033-1047

Trappe M (1982) Verhalten und Echoortung der Grossen Hufeisennase beim Insektenfang. Thesis, University of Tübingen, FRGermany

Vater M (1981) Single unit responses to linear frequency-modulations in the inferior colliculus of the Greater Horseshoe bat, Rhinolophus ferrumequinum. J Comp Physiol 141:249-264

Vater M, Schlegel P (1979) Comparative auditory neurophysiology of the inferior colliculus of two Molossid bats, Molossus ater and Molossus molossus. II. Single Unit responses to frequency-modulated signals and signal and noise combinations. J Comp Physiol 131:147-160

Webster FA (1963) Active energy radiating systems: the bat and ultrasonic principles II, acoustical control of airborne interceptions by bats. Proc Int Congr Tech and Blindness A F B, New York 1:49-135

Time Constants of Various Parts of the Human Auditory System and Some of Their Consequences

Per V. Brüel and Keld Baden-Kristensen
Brüel & Kjær, DK 2850 Nærum, Denmark

INTRODUCTION

Noise, according to a rather hackneyed expression, is undesired sound; that is, sounds that disturb, annoy, and even impair hearing. Nevertheless, the internationally standardized Sound Level Meter (IEC 1973) has been developed entirely on the basis of arbitrarily agreed equal loudness contours (inverted), without due consideration of the sounds that disturb or annoy, and (which is much worse) none at all to those which involve a risk for hearing damage. When one therefore asks, Do we measure noise correctly? the answer must be that where hearing level is concerned, the scale in use today is applicable because it was originally developed on the basis of hearing level. If, however, one considers the annoyance caused by noise, then our noise scale is no longer appropriate, and even less acceptable when used to stipulate permissible noise limits to prevent hearing loss. The latter is rather serious, since large sums are offered for prevention of hearing loss caused by industry and traffic.

PSYCHOACOUSTIC DATA RELEVANT FOR SOUND LEVEL METERS

The loudness of a sound perceived by the ear is a function of amplitude and the duration of the stimulus presented to the ear. Under some circumstances, continual growth in loudness for a stimulus with constant amplitude may occur for up to 1 second. If, on the other hand, the same stimulus existed for a shorter time (less than 200 ms), there would be a reduction in loudness, and the shorter the duration of the stimulus, the less loud it would be.

Several researchers have investigated this phenomenon of temporal integration of loudness, and many of their results are shown in Fig. 1. The results were obtained by psychoacoustic experiments in which the loudness of impulsive stimuli such as tone or noise pulses of different duration were compared to those of steady sounds. The sound pressure level difference (L_i-L_D), in dB, between the impulse level L_i and that of the steady sound L_D, which was judged to be equally loud, is used as ordinate in the figure. As impulse level L_i, the level of the signal from which the pulse was cut out was used. The x-axis shows the duration of the pulse. As long as the duration of the pulse is longer than a certain amount, the sound pressure level of the pulse and that of the steady sound are equal for equal loudness judgment: (L_i-L_D) = 0. However, for

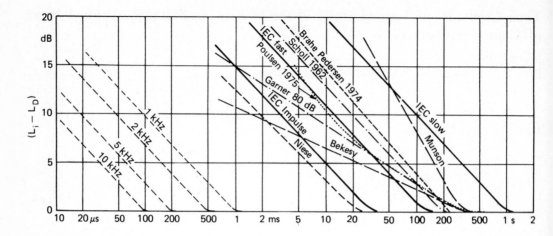

Fig. 1. Results form different researchers of the subjective perception of short impulses compared with the integration curves for the
time constants "Fast", "Slow" and "Impulse" of the sound level meter.
To the left are shown derived integration curves for cochlear filters
at high SPL.

pulses shorter than this duration, the level of the pulse must be
increased to give the same loudness sensation as the steady sound.
The breaking point at $(L_i - L_D) = 0$ then corresponds to the effective
integration time of the ear.

A considerable spread can be seen in the data of Fig. 1. Differences in absolute level and frequency content of the stimuli as
well as differences in experimental paradigms may be responsible.
Based on such data the International Electrotechnical Commission
(IEC) has chosen detector time constants of 100 ms (fast) and 1 s
(slow) for sound level meters. Furthermore, short impulses having
equal energy should yield the same meter deflection.

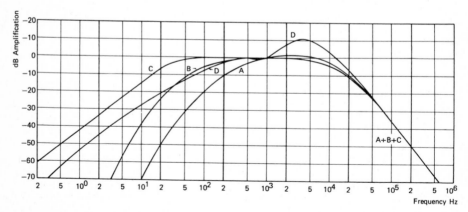

Fig. 2. The weighting curves of the IEC standardized sound level
meter. A, B and C curves originally the reciprocal of Fletcher and
Munson's curves.

The frequency sensitivity of the human ear was originally determined by Fletcher and Munson (1933), and later modified slightly by
Churcher and King (1937). From these equal loudness contours, the
weighting curves for the sound level meters have been standardized
by IEC (1973), and are shown in Fig. 2. From the slope of the
curves at high frequencies, it can be seen that the upper frequency
limit corresponds to a double RC network each with a time constant
of 13 μs. As there are two RC networks in series the pulse response
is not a simple exponential function but a combined one with an
equivalent time constant of approximately 35 μs.

Whereas the IEC sound level meter by definition is doing a reasonably good job in assessing loudness, this cannot be the case when
trying to use it to assess risk of noise induced hearing loss, in
particular when highly impulsive noises are involved.

In order to support that statement two unsettled issues have to be
dealt with:
i) We must try to extrapolate or infer, in part from animal data,
how the peripheral auditory system, subject to damage, will function at very high sound pressures and
ii) we must make a statement about what (assumed) aspects of the
sound stimulus generate irreversible hearing loss.

Then, by comparing with the known characteristics of the IEC sound
level meter, the inadequacies will be apparent.

CHARACTERISTICS OF THE PERIPHERAL AUDITORY SYSTEM AT HIGH SOUND
PRESSURE LEVELS

A schematic drawing of the auditory system emphasizing the periphery is shown in Fig. 3. Also, the frequency and time characteristics of the outer and middle ear and of the basilar membrane portion of the inner ear are shown.

The outer ear exhibits a wide frequency response with a broad quar-
ter-wave resonance peak around 4 kHz (Wiener and Ross 1946). The
middle ear also exhibits broad frequency response without sharp
resonances. So, the combined frequency response of the outer and
middle ear is broad, still with the quarter wave resonance at 4 kHz
present.

The combined frequency response is to a large extent responsible
for the all over frequency sensitivity of the ear (Lynch et al.
1982), reflected in equal loudness contours and weighting curves,
Fig. 2. Actually, the D curve more closely reflects the combined
response than the A curve in the 1-8 kHz range, since the 4 kHz
peak region is better approximated. Since the frequency weighting
curves of the outer and middle ear and IEC sound level meters are
closely related, so are their respective filter time constants.
Looking closely, the time constant of the IEC weighting curves, 35
μs, is only slightly smaller than the approximately 50 μs of the
outer and middle ear. Apart from the low frequency attenuation
resulting from contraction of the stapedius muscle, the outer and

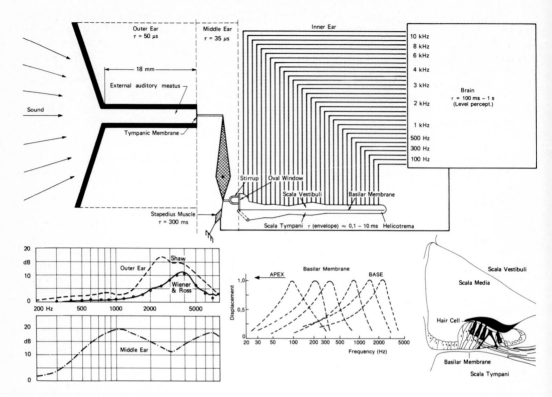

Fig. 3. Schematic drawing of the human ear with the system's most important time constants, frequency curves, frequency positions and integration constants.

middle ear transmission system has a linear amplitude response up to about 140 dB SPL (Guinan and Peake 1967), i.e. also for damaging noise levels. At even higher levels stretching of ossicular joints and ligaments sets in.

Having reached the stirrup, the noise signal is transmitted into the inner ear, the cochlea, for further filtering and detection. The stirrup motion generates pressures in the cochlear fluids, which act as driving forces on the basilar membrane and the associated transducer (hair) cells and tectorial membrane, shown black in the lower right insert, Fig. 3. As a result the basilar membrane and the sensory hairs are deflected. This deflection is the key mechanical excitation of the transducer cells, which initiates a chain of events ultimately leading to auditory nerve firing and transmission to the brain. As a result of a varying mechanical stiffness along the length of the basilar membrane, each location and thus subgroup of hair cells and nerve fibers have a range of "best frequencies" to which they are most sensitive (Békésy 1960; Siebert 1974; Russell and Sellick 1978; Kiang 1965), see Fig. 3.

The frequency selectivity of individual cochlear nerve fibers in the frequency-region around the best frequency has long been known to depend on the best frequency (Kiang 1965) as well as on stimulus level (Anderson et al. 1971). For sound pressure levels close to 0 dB high Q values are found for cochlear filters: 15 at 10 kHz, 3 at 1 kHz are typical values. As stimulus level is increased, Q values decrease corresponding to a larger effective filter bandwith and indicating an increase in the filter responsiveness in the time domain. Level dependence of frequency selectivity is also found in the response of individual hair cells (Russell and Selick 1978) and in the response of the basilar membrane (Selick et al. 1982), see Fig. 4.

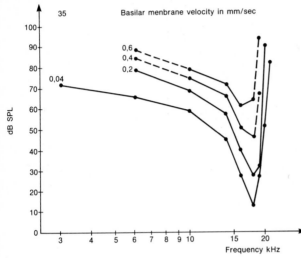

Fig. 4. Basilar membrane iso-velocity curves based on data from Sellick et al. (1982), Figures 5a and 10. Full lines are drawn between experimental or interpolated datapoints. Broken lines includes an extra-polated datapoint. The frequency selectivity in the region of most sensitive frequencies is decreasing with increasing SPL.

Since hearing damage is at issue, filter characteristics at 120 dB SPL and upwards are of interest. No recent, systematic data of that kind are available, however von Békésy's classical data on cadavers (1949) probably yield the right order of magnitude. Using his data and converting effective bandwidths to approximate filter time constants, the broken lines of Fig. 1 left are arrived at. They will slide along the abscissa for changing SPL and should be used cautiously. Even so, it is quite apparent that impulsive stimuli presented at high levels need very short (envelope) durations for the cochlear filters to arrive at full vibration amplitude, corre-sponding to that of continuous SPL excitation.

ASSESSMENT OF RISK OF HEARING LOSS GIVEN ASSUMPTION OF LOSS MECHANISM

If we assume that the amplitude of mechanical excitation of the transducer cells is a key variable in loss generation, then it is obvious that sound level meters employing integration times of hundred to thousand milliseconds cannot possibly assess risk of damage occurring in the one millisecond range. The underestimate could easily be 20 to 30 dB depending on the spectral content of the impulse and its duration.

This in fact explains the paradox, that practically all noise induced hearing loss starts in the frequency range around 4 kHz,
inspite of the fact that most industrial noises, vehicle noise as
well as noise from lawn mowers, music from radios and in dis-
cotheques have their largest components in the frequency range
100-2000 Hz. Only a small portion of the total sound energy lies in
the 3 kHz-10 kHz range, see Fig. 5. The spectra illustrate the

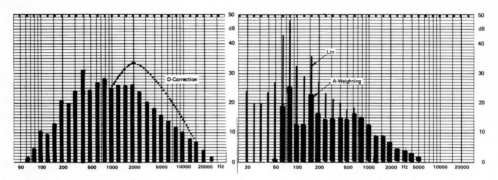

Fig. 5. Left: 1/3 octave spectrum of noise from a typical metal
industry. Rigth: Spectral distribution of noise in a propeller driven
aircraft.

frequency distribution of the sound we hear, i.e. the energy distribution integrated over at least 100 ms. On the other hand, if we
examine single impulses which are short and widely separated in
time, a completely different picture is obtained. For example,
noise from hammer blows on brass and steel respectively, give a
ringing sound as can be seen in Fig. 6, and contain considerable

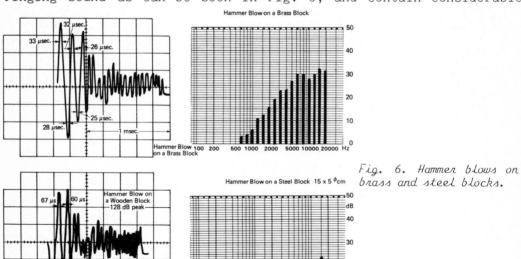

Fig. 6. Hammer blows on
brass and steel blocks.

energy in the frequency range 4 kHz–16 kHz. The frequency spectra are obtained from an analyzer covering a wide frequency range and show that the bulk of the energy is radiated in less than 500 μs.

From Fig. 1, using for example IEC Fast, it can be seen that the short hammer blows will be measured and also give a subjective impression approximately 25 dB lower than the peak output amplitude of the cochlear filters most sensitive to the frequency content of the blow. It is therefore possible to have a workshop in which the equivalent noise level is for example 90 dB(A), whilst hammer blows there can momentarily give sound levels up to 110-115 dB(A). These peaks have the largest energy content precisely in the same frequency range where induced hearing loss is found. If the ear is in close vicinity of the hammer blows, very high sound levels would be experienced. Sound impulses from blows are often found to have peak values up to 145 dB. There are indications that the induced hearing losses are primarily caused by the very short sound impulses, which are commonly found in noisy industrial environments, and to a less extent by a high average noise level. Fig. 7 shows noise in a me-

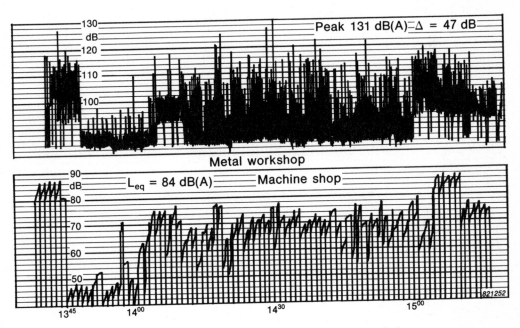

Fig. 7. Recording of peak levels and L_{eq} in a metal workshop.

chanical workshop measured synchronously by a sound level meter, that measures average sound energy every minute (L_{eq}), and by an-

other sound level meter with a time constant of 30 μs, which can measure up to three times a second the maximum sound level in the foregoing 1/3 of a second. One can see here a few peak levels which are 47 dB higher than the average sound level. These noise histories are typical for a workshop where metal parts are manufactured, and the short high sound levels are caused by metals striking each other, for example, by hammer blows, punch presses, riveting machines etc.

In wood industries where metals strike wood, the peak levels are less high, but on the other hand the impulses are longer. The average noise level in wood industries is often higher than in a mechanical workshop, but as a rule the noise peaks are lower (see Fig. 8). This explains the apparent paradox that workers in the

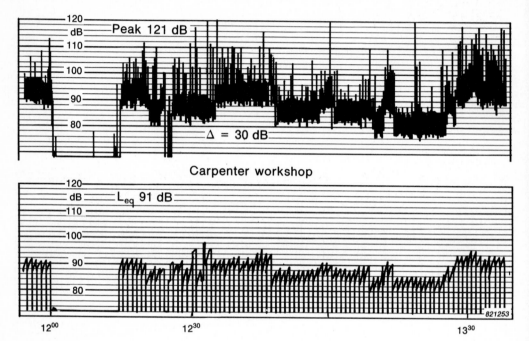

Fig. 8. Recording of peak levels and L_{eq} in a carpenter's workshop.

wood industry can tolerate on an average 10-15 dB higher levels than workers in the metal industry. This surprising result was revealed by an extensive Dutch survey (Passchier-Vermeer 1971), where the induced hearing loss was correlated with the average noise levels in different industries. If the induced hearing loss is caused primarily by the noise peaks, this result is quite obvious.

Another investigation shows that approximately 25% of all male inhabitants in Denmark and Sweden, even in their young age, have a hearing loss at 4 kHz; as a rule it is not serious and most are also unaware of this handicap. The reason why so many young males have this hearing loss probably is that they have been in close vicinity of a fire cracker or a gun shot from a light weapon. If it is argued that peak levels of such impulsive sounds above about 155 dB do result in permanent damage (Price 1984), then the 170 to 175 dB, which is easily reached, will create permanent damage, Fig. 9, even though the impulses can be so short that they do not sound so loud. The oscillograms from a small signal pistol and a widely

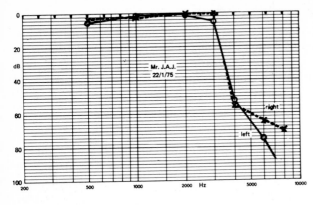

Fig. 9. Typical audiogram illustrating permanent hearing loss of a person who has been exposed to gunshot.

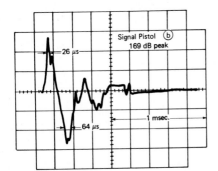
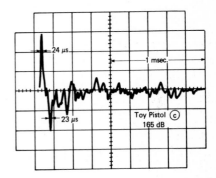

Fig. 10. Oscillograms of a shot from a signal pistol at 1 m distance and of a shot from a toy pistol with cap at 10 cm distance.

available toy pistol, Fig. 10, display peaks of 169 and 165 dB respectively, which indicates that even these may be capable of generating permanent damage.

DISCUSSION AND CONCLUSION

The present article deals with time constants of cochlear filters extrapolated to values for high SPL excitation relevant to hearing damage. Assuming that the peak output amplitude of these filters is a key variable in generating hearing loss, the consequences of the millisecond range time constants for risk assessment are looked at. The fact that effective bandwidths and associated time constants of cochlear filters are strongly dependent on excitation level means that the simple "mechanical filter" approach used here is an approximation. Factors such as state of adaptation and time dynamics of change of state ultimately need to be dealt with. The time history preceeding individual impulses, known to affect loss (Hamernik et al. 1980) might then be accounted for.

Longer time events than treated here are also necessary in an ultimate complete picture. They are certainly associated with the state
of the whole cochlea including biochemical, vascular, or tissue
mechanical (swelling and tissue repair) factors.

Based on our present knowledge of time "constants" of the ear and
of the occurrence of high level sound impulses in our daily life it
seems safe to conclude, that our criteria for dangerous noise
levels should be revised, and the noise limits in our industries
should not solely be based on the average sound levels (Price 1984).

REFERENCES

Anderson DJ, Rose JE, Hind JE, Brugge JF (1971) Temporal position
of discharge in single auditory nerve fibers within the cycle of
a sine-wave stimulus: Frequency and intensity effects. J Acoust
Soc Am 49:1131-1139
Békésy G von (1949) On the resonance curve and the decay period at
various points on the cochlear partition. J Acoust Soc Amer
21:245-254
Brüel PV (1977) Do we measure damaging noise correctly? Noise Control Engineering, March/April
Churcher BG, King AJ (1937) The performance of noise meters in
terms of primary standard. Proc Inst Electr Engrs. England,
London, 57-90
Fletcher H, Munson WA (1933) Loudness, its definition, measurement
and calculation. J Acoust Soc Amer 5:82-108
Guinan JJ Jr, Peake WT (1967) Middle ear characteristics of ane-
sthetised cats. J Acoust Soc Amer 41:1237-1261
Hamernik RP, Henderson D, Salvi RJ (1980) Contribution of animal
studies to our understanding of impulse noise induced hearing
loss. Scand Audiol Suppl 12:128-146
IEC "International Electronical Commission", Geneve, Publication
179: Precision Sound Level Meters (1973) and revised edition 651
(1980)
Kiang NYS (1965) Discharge patterns of single fibers in the cat's
auditory nerve. MIT research monograph no 35, chapter 7. MIT
Press, Cambridge, Massachusetts
Lynch TJ III, Nedzelnitsky V, Peake WT (1982) Input impedance of
the cochlea in cat. J Acoust Soc Amer 72:108-130
Passchier-Vermeer W (1971) Steady-state and fluctuating noise. Its
effects on the hearing of people. Occupational hearing loss.
British Acoustical Society. Special Volume no 1, London, p 15
Price GR (1984) Practical applications of basic research on impulse
noise hazard. In: Proceedings of the 1984 International Confe-
rence on Noise Control Engineering, Hawaii, 821-826
Sellick PM, Patuzzi R, Johnstone BM (1982) Measurement of basilar
membrane motion in the guinea pig using the Mössbauer technique.
J Acoust Soc Amer 72:131-141
Russell IJ, Sellick PM (1978) Intracellular studies of hair cells
in the mammalian cochlea. J Physiol 284:261-290
Siebert WM (1974) Ranke revisited - a simple short-wave cochlear
model. J Acoust Soc Amer 56:594-600
Wiener FM, Ross DA (1946) The pressure distribution in the auditory
canal in a progressive sound field. J Acoust Soc Amer 18:401-408

Temporal Patterning in Speech: The Implications of Temporal Resolution and Signal-Processing

Mark Haggard

MRC Institute of Hearing Research, Nottingham NG7 2RD, England

INTRODUCTION

There are wide variations in behavioural indices of temporal resolution in humans, with degradation of resolution accompanying both brain damage (Lackner and Teuber 1973), and hearing impairment (Tyler et al. 1982). The presence of only minimal abnormalities of temporal processing in animal models of cochlear pathology (Harrison and Evans 1979; Evans 1984, and papers cited therein) raises some doubt as to whether degradation of temporal resolution is a direct consequence of cochlear pathology, or some loosely associated correlate, or the result of abnormal central neural organisation following the lack of afferent stimulation. While physiological answers are being sought to such questions, it is profitable (i) to develop the ecological description of temporal structure in significant auditory events; (ii) to develop an understanding of the perceptual processes and phenomena resulting from temporal structure; and (iii) to develop ways of transforming that structure which are scientifically and technologically useful. This paper is directed to those three ends. We must remember, however, that the contrast between time- and frequency-based structures is more a dichotomy of our explanatory framework than a dichotomy in the reality we seek to describe. Auditory events are <u>spectro</u>temporal, and the reason for concentrating now on the temporal aspect is merely to redress a historical imbalance.

NECESSARY TEMPORAL DESCRIPTORS OF STIMULUS STRUCTURE

Sound waves result from energy sources and they propagate and reflect as waves within conducting media; the main determinants of audio frequencies, f, are the sound propagation velocity in the medium and the dimensions of the object or of the spaces containing the medium enclosed by a more rigid medium. These sound waves are able to carry reliable information about the detailed time-history of <u>changes</u> in the objects or the medium, and in their dimensions, only if the time-course of such changes is somewhat longer than the period of the audio (carrier) frequencies concerned. On sampling theory, given an ideal filter, only twice as many digital samples need be taken per second as the highest frequency transmitted, but in real biological systems a large safety margin is provided by Nature so that central processing only attempts problems where it will not be slave to peripheral constraints. With a safety factor of 100 for the ratio of event time course to carrier period, events repeating on the order of 5 to 40 per second will lie within the range for which audio frequencies of 500 Hz to 4 kHz respectively (the region of maximum auditory sensitivity and resolution by cochlear mechanics in the human) can comfortably act as carriers. Special perceptual mechanisms exist for extracting periodicity pitch from complex tones, especially those having lower fundamental periodicities than about 500 Hz, but the register of the human bass voice from about 50 Hz upwards defines an audio periodicity range into which the slower fluctuations under discussion here cannot be considered to encroach. We may therefore expect the auditory system as a whole to tailor its best absolute and differential response to auditory events with repetition rates, F, of up to about 40 cycles/sec; that is, with (half-cycle) single events having durations or transitions down to about 1/80 of a second (12 ms), as found in speech. Separate mechanisms for handling changes outside this time scale may have evolved; for example in vision there seems to be a system for handling slow changes in luminance separate from that for handling flicker as such (Kelly 1971).

Modulation terminology and measurement

To describe changing acoustical events we need the concept of low modulation frequency. To avoid confusion I will denote this as F in cycles/sec rather than f in Hz, and will describe F-regions as "sub-ranges" rather than as bands. The neologism "cepstrum" was coined to refer to the spectrum of the log power spectrum, and this in turn baptised the associated operation of "liftering". I feel justified in coining the word "tilfering" for filtering the modulation spectrum because the concept is relatively new; but I am reluctant to force "pecstrum" on the reader when the term "modulation spectrum" is already used. There will in general not be any f-spectral component in the F-range unless there is non-linear distortion, and it would in any case be largely "infrasonic", i.e. inaudible.

We can divide up the auditory frequency spectrum into a number of narrow bands, and then concentrate on the (rectified) power changes in each band. Structured sounds such as speech, involving spectral changes at various rates, must have modulation spectra spanning the F range; the power in this spectrum is better thought of statistically, as variance rather than energy. The modulation spectrum is essentially statistical and need not always provide the most useful description. It may be simplest in some circumstances to say that intensity increased at time t, and reduced at time t_2, or that an interval of duration (t_2-t_1) was present in the signal around time t. The alternative is to say that its modulation spectrum has a peak at zero, and at $F = \frac{1}{2(t_2-t_1)}$ and its odd multiples, with a null at $F = \frac{1}{(t_2-t_1)}$ and integer multiples.

Recent years have seen a growth of the use of temporal modulation transfer functions (TMTF) in the description of the temporal resolution properties of physiological and engineering systems. The TMTF is a scaled ratio between modulation spectral components at input and output, usually expressed logarithmically as a difference in decibels. For a formal treatment, see Schroeder (1981). Absolute modulation spectra themselves have so far been used very little as specifications of auditory stimuli. It is not my intention to proselytise modulation spectra as they may not always be needed. For example when discussing the auditory difference between a bottle breaking and bouncing, Warren and Verbrugge (1984) found it sufficient to move directly from spectrograms to a characterisation in terms of presence/absence and duration of burst and silence portions of the waveform envelope. There was some degree of period-icity in the burst/silence alternations for bouncing, which might encourage description of the contrast in terms of peakedness of the modulation spectrum, or in terms of autocorrelation in the ampli-tude envelope. But such descriptions will have to be evaluated as useful or otherwise once auditory ecologists become aware of them.

One disincentive to specifying modulation spectra is the computatio-nal inconvenience of deriving them at present. Hardware or software must first derive a succession of audio-frequency spectra, then operate upon the scaled power measurements for particular frequency bands to produce spectra of those abstract signals composed by considering each band's successive level values as a time series. In a stimulus such as speech, in which the main component frequen-cies are continually changing, the single modulation spectrum for the time-envelope of the entire broadband (f) signal will grossly underestimate the amount of informative dynamic change, therefore the data of Fig. 1 are based on narrow bands. A complete charac-terisation would require some 20 modulation spectra, one for each

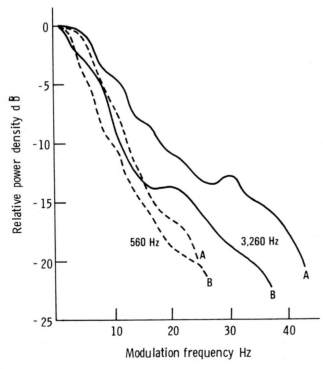

Fig. 1. Modulation spectra for a high-frequency band and a low-frequency band (bandwidth = 180 Hz in each case) for two adult male speakers, replotted from Holmes (1961). Note the linear F-scale.

of 20 contiguous audio frequency bands; in providing this, Steeneken and Houtgast (1983) have shown that the bands' modulation spectra have much in common. The number 20 is about the number of discrete bands required to approximate speech sound spectra, considering the bandwidths of speech formants; one twentieth of the width of the speech spectrum (on a logarithmic or critical band representation) should therefore reflect the important power fluctuations.

There is not space to treat here the sub-problem met when the auditory resolution bands are narrow (at low f), yet the harmonics (e.g. of a female or child voice source) are widely spaced; here changing fundamental frequency will introduce a further source of variance into the intensity signal for the band, that will need to be discounted in extracting those variance components attributable to the formant movements.

Statistical role of modulation in communication

The work of Steeneken and Houtgast (1980; see also Plomp 1983) has done much to establish the utility of the modulation spectrum and modulation transfer function as a statistical descriptor of information transmission. Briefly, they showed that many types of speech interference that appear heterogeneous when specified in waveform or audio spectrum terms can be reduced in their intelligibility-degrading effect, to a single common scale reflecting loss of (spectro)temporal definition; this is conveniently summarised in the TMTF. Because of the similarities in the long-term spectra of different bands, the modulation spectrum for very few such bands, (e.g. two, centred on 0.5 and 2.0 kHz) appear to provide a sufficient statistical description of general signal quality in typical applications. The limited modulation frequency range (16 cycles/sec) used with success would suggest that the modulation frequencies most important for intelligibility do not extend much higher than this. Thus Steeneken and Houtgast have demonstrated that modulation measurements need not be comprehensive to be useful.

Steeneken and Houtgast's work suggested that reduction of intelligibility is very closely and causally related to the reduction of mean temporal modulation transfer when measured on narrow bands. Effects of white noise addition can be represented as a reduction of modulation transfer due to a clamping by the noise's mean value, the noise's power variance only becoming material for very narrow bandwidths. The envelope-coherent distortion from non-linear circuits introduces components that tend to flatten the spectrum within any short time sample and hence to depress temporal modulations at any frequency. Lundin (1982) notes that room reverberations depress the low-F modulations below 4 cycles/sec but add appreciable spurious energy above 16 cycles/sec, with a merging region between. The spurious addition is probably due to harmonics and formants rapidly crossing the room resonances. (Now we know why plainchant evolved in mediaeval cathedrals - to improve intelligibility by minimising the non-phonetic sources of modulation produced by resonance-crossing). The generation of spurious high-F modulations suggests a class of degradation for which Steeneken and Houtgast's TMTF formulation for intelligibility would need to be modified. When the interference itself has a modulation spectrum comparable to that of the speech, or when the distortion raises levels in the modulation spectrum, obviously, the TMTF becomes positive despite intelligibility decreasing. Here the more general metric of correlation of all the particular input band modulation patterns with their corresponding output modulation patterns would relate more fundamentally to intelligibility, but there would be some loss of predictive simplicity in the method.

No precise determinations of the modulation spectrum's F-importance function for speech intelligibility have been made, but some pointers exist. There is agreement on speech parameter digital sampling rates in analysis-synthesis telephony (e.g. formant synthesisers, and vocoders), that about 50 samples/sec is adequate for most purposes. By invoking a Nyquist factor of just over 2 this places the important range below about 20 cycles/sec. The majority of important segmentally-related modulation information has to lie in the two octave F-ranges between 4 or 5 cycles/sec (roughly the syllabic rate) and this limit. This is broadly supported by the distribution of the typical durations of important phonetic segments in speech. Both Holmes (1961) and Steeneken and Houtgast (1983) show that the modulation spectrum is rather steeply sloping; the considerable energy (i.e. variance) below about 4 cycles/sec transmits information about overall amplitude and source characteristics associated with syllable accentuation and sentence constituent structure, but cannot contribute to word intelligibility except indirectly. Circuits in that very broad class known as amplitude compressors remove some of this common slowly varying intensity fluctuation, with virtually no loss of intelligibility and indeed they are noticed more by their raising of the background noise level than by their effect upon signal quality or intelligibility. (See also the later section on "Spectral approaches to spectro-temporal enhancement"). The information important for intelligibility lies therefore in a rather narrow range, the middle two sub-ranges of the total range under consideration, i.e. from 4 to 8 and 8 to 16 cycles/sec where not all the fluctuations are common to all f-bands.

Our appreciation of the origin and role of the modulation spectrum can be structured by appeal to universal physical constraints. Despite the necessity for distinctions of terminology, analogies from the more familiar and primary audio f-spectrum can assist us in thinking about the modulation spectrum. Because of basic physical constraints in the larynx the speech f-spectrum slopes at between -6 dB and -12 dB per octave between 0.5 and 4.0 kHz. For information-theoretic reasons it is efficient in transmission to lift the spectrum by about +6 dB per octave, i.e. roughly to flatten it; the effects upon intelligibility are generally worthwhile, but particularly so if the listener already has a hearing impairment or if the speech is to suffer additive broadband noise in transmission. For analogous reasons of physical and physiological inertia in speech production, the modulation spectrum also has a slope in the -6 to -12 dB per octave range, as can be inferred from Fig. 1, and more directly from Steeneken and Houtgast's data (1983).

For efficient transmission, should we analogously tilfer the modulation spectrum, by +6 to +12 dB per octave, especially as the sub-ranges below 4 cycles/sec appear to contribute little to segmental aspects of intelligibility? One way to do this would be to introduce first- or second-order differentiation with respect to time of the intensity signals in the various audio-frequency bands. Langhans and Strube (1982) tilfered short-term modulation spectra in several ways. One involved performing an LPC analysis, non-linear transformation and resynthesis procedure; they reported enhancing effects on speech transmission through added noise. Their result suggests that the middle sub-ranges above 4 cycles/sec carry much of the information important for intelligibility. The intelligibility of speech should not suffer greatly if the amount of modulation were enhanced or suppressed in a narrow band of F-frequencies, so long as F phase-shift (i.e. temporal dispersion) is minimised, and so long as overall contrastive patterns of low-F and high-F modulation are maintained.

Physical-phonetic interpretation of F

Each class of speech sound will have a continuous modulation spectrum, though in practice one would be more likely to calculate long-term modulation spectra, as the basic concept is inherently statistical. With the possible exception of trills, as in Spanish and Scottish /rr/, individual speech sounds do not have marked periodicity in their envelopes in the f range below the F of the voice fundamental and thus do not have highly peaked F-spectra. Hence, as with bands in the f-spectrum, the different modulation frequency regions within the important part of the F-spectrum cannot be associated exclusively or even closely with particular phonetic values of speech sound features. However it is obvious that sounds with abrupt amplitude changes, such as stops and affricates, must contribute more to the upper sub-ranges of F, while the slowly changing vowels, glides and diphthongs must contribute more to the lower sub-range. Due to the rapidity with which the aerodynamics of turbulence permit sources forward in the vocal tract, as in fricatives and plosives, to commence and cease generation, coupled with their characteristic audio spectra, this implicates particularly the high-f region; there is a little evidence for this in the high-F bumps in the higher-f band in Figure 1. However, in general F-information will be independent of f.

I have made some use of an analogy between F and f for explanatory purposes; it holds for basic reasons of physical constraint and information theory. The scope for explanatory analogy may reach its limit in the question as to whether separate F-bands can be considered separate analysers, or functional channels. This question has two forms, not identical. Firstly the information in separate bands might combine in systematic ways, eg the mutual redundancy of bands might decrease continuously with their separation in F-space; this is a question about the statistical ecology of speech with a likely positive answer. Secondly the analysis of adjacent bands of a certain width might be functionally, and to an extent physiologically, separated; such separation occurs with the auditory critical band and the neural single-unit characteristic in f-space, and its peripheral biophysical basis is now quite well understood (Evans 1978). To attribute to the auditory nervous system analysers tuned to narrow ranges of F values is a much more risky hypothesis. There is no direct evidence for this; evidence from perstimulatory adaptation experiments such as those by Tansley and Suffield (1983) is adequately interpreted as reflecting the F-response of the auditory system's registration of modulation as a whole. However physiological evidence for cortical units selective as to rate of modulation entails that narrowly F-tuned modulation analysers cannot be ruled out.

TEMPORAL MODULATION IN HEARING AND AUDITORY PATTERN PERCEPTION

The use made of temporal patterning in perceptual processing by the brain can be illuminated in many different types of experiment. In this section I wish merely to review three types of experiment which together give us a picture of the role of primarily temporal factors and how they could be affected by loss of temporal resolution.

Correlates of temporal resolution

Tyler et al (1982) sought to test the precise role of temporal degradation in impaired hearing of cochlear origin by predicting with partial correlations a set of hearing-impaired listeners' scores in identifying words in speech-shaped noise. They found that measures of frequency resolution and of absolute auditory sensitivity, being somewhat closely correlated, represented a single factor in accounting for the scores. In other words, after partialling the effects of either one, the other did not add significantly to the prediction. However, independent of whichever was first partialled, the further incorporation of the gap detection threshold (a measure of temporal resolution) did significantly enhance the prediction of word identification scores in noise, summarising an independent contributory influence on auditory disability.

These apparently independent contributions of frequency resolution and temporal resolution have had a material influence upon our subsequent work on auditory disability, arguing for incorporation of measures of temporal resolution as well as frequency resolution in clinical studies. The pattern of correlations observed is theoretically acceptable at the most general level. Loss of frequency selectivity and of absolute sensitivity have a common physiological basis (Evans 1978). A broadening of auditory analyser bandwidth beyond the physical bandwidth of speech formants would lead to some reduction of the modulation energy in each auditory f-band, due to spectral averaging. This reduction would be particularly marked for those middle/high-F components characterising formant movements, although not necessarily for the high-F components of abrupt onsets in broad-spectrum sounds such as fricatives, stops and affricates. So some reduction of the effective temporal modulation transfer for spectrotemporal changes in speech could be expected from loss of frequency resolution alone, but degradation of broadband gap-detection cannot be explained in this way. Also the absence of sufficient correlation between the frequency and temporal measures to preempt the latter's predictive contribution argues that this account is insufficient and that the physiological basis of this particular temporal measure's variation lies elsewhere, e.g. centrally. It was noted earlier that there were no gross abnormalities of temporal representation in cochlear pathology, but before a central effect is assumed, there should be further examinations of peripheral coding in animal models of pathology, in particular the time-course of recovery from post-stimulus suppression (Evans 1984).

A range of alternative hypotheses, graded in specificity, is to hand. At one interesting extreme auditory deprivation could lead to desynchronisation of central neural firing; this is possible although not very likely given evidence of no degradation of synchronisation in developmentally sound-deprived mice (Horner and Bock 1984). At the other less interesting extreme, people of least ability or motivation could be those scoring worst on both the word identification task and the gap detection task; this is also unlikely as Tyler et al did not find a prominent general factor (no high intercorrelation of test results) nor was the correlation between speech-identification and gap-detection much higher than that between random pairs of test results. In the absence of direct evidence an intermediate hypothesis must be sought, consistent with the absolute magnitude of the duration thresholds for gap detection. These were of the order of 12 ms (with a wide variance) for the impaired group and 7 ms for the controls. These durations are of the same order as those found in tasks affected by damage to the cerebral cortex (Lackner and Teuber 1973). They are a little too long to be plausibly ascribed directly to abnormality in peripheral processes of frequency resolution where the prediction from a resonance model would in any case be for an improvement in time resolution; but they are too short directly to influence the phonetic distinctions that are most obviously based upon gap detection. The half-cycle duration at the upper effective limit of the modulation spectrum (50 cycles/sec), is 10 msec, exactly in the zone that distinguishes the hearing-impaired in gap detection. Degradation in this range is likely to have only marginal <u>direct</u> effects upon speech perception. However these gap-detection thresholds could reflect a general degradation of temporal processing (Bailey 1983) that has other subtle phonetic perceptual effects we do not yet understand.

One likely class of explanation is that correlates of hearing disorders such as age or cardio-vascular disease also underlie the loss of this form of temporal resolution. Shailer and Moore (1983) establish that high-f channels have a dominant role in temporal resolution; Buus and Florentine (1984) and Henderson et al (1984) show respectively in humans and in animals affected by high frequency hearing loss that the two types of resolution deficit are associated. These are the results we would expect on time-frequency reciprocity from the broad analyser bandwidth of the basal portion of the cochlea but they do not establish that the whole temporal processing deficit of an impaired group or the variation within that group must be ascribed predominantly to the loss of high-frequency sensitivity. It is worth looking to other sites that could be influenced in some types or degrees of pathology, such as the synapse, or even beyond the cochlea.

Peripheral/Central Division of Labour

Summerfield et al (1983) produced auditory demonstrations of a successive contrast effect which appears to be a consequence of internal physiological sharpening of temporal representations, partly analogous to the desired flattening of the F-spectrum by tilfering. They showed that a buzz with a uniform f-spectrum could on occasions be heard as vowel-like despite the absence of appropriate formants. This occurred when it was preceded by a sound having valleys at frequencies where the following buzz should otherwise have had to possess formants to be heard as a vowel. The temporal contrast effect can be explained by peripheral short-term adaptation - presumably at the hair cell's synapse. Delgutte and Kiang (1984) have shown in a convincing series of animal investigations that auditory nerve representations of phonetically relevant properties of speech sounds are indeed modified by adaptation phenomena, apparently in ways beneficial for communication.

Adaptation does not appear to be the whole story. Summerfield et al found temporal enhancement phenomenon such as peripheral adaptation to be a necessary part of but not the whole explanation of their results. Their contrast effect could be demonstrated for proceeding dips as shallow as 2 dB and for valley-buzz separations as long as 100 ms. This suggests that the phonetic interpretation of the spectrum is assisted not only by enhancement of the auditory nerve representation over the order of a few tens of milliseconds, but also by central (memory-based) comparisons of a contrastive nature. There is evidence that analogous after-effects can be sustained for many hundreds of milliseconds with non-speech sounds (Wilson 1970; Viemeister 1980), arguing for a central sequential contrast with a spectral memory. There is an obvious evolutionary utility in detecting change against the relatively static background of spectra such as those of sea and wind; the new event may signal prey or danger. Likewise irregular f-transfer functions conditioned by electroacoustic equipment, or by wavefront propagation in and near structures such as human heads and in buildings, make absolute spectral magnitudes unreliable. But these properties are relatively static, allowing the brain to separate out the patterns of change that can be ascribed to movement of articulators or movements in space. We are forced to the conclusion that the biological burden of temporal pattern extraction is shared between ear and brain.

Grouping by modulation characteristics

The ability to perceive triple spectral amplitude increases as equivalent to three static formants of a vowel in the Summerfield et al experiment, must depend upon grouping of the components according to their common amplitude trajectory; the common fundamental, otherwise a typical basis of grouping, could not serve in their experiment to differentiate the critical components from the other non-increasing components in the uniform buzz spectrum. Darwin and Sutherland (1984) have likewise demonstrated that the common amplitude trajectory of harmonics is an important determinant of whether or not they are perceptually included in the representation of a source, as inferred by whether or not they contribute to its phonetic quality. This temporal grouping by

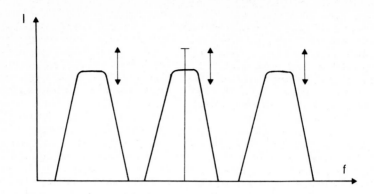

Fig. 2. A spectral representation of the phenomenon of comodulation masking release (CMR), which occurs when intensity fluctuations, symbolised as vertical arrows, take the same form in the critical band round a probe tone and in a flanking band.

common amplitude modulation is one of many principles of grouping long known to composers (McAdams 1984). The question then arises: does this grouping apply only to grouping of suprathreshold sounds into subjective sources? Or is it a more general principle that applies also to masked thresholds, and hence stands to interplay with the various factors affecting masked thresholds, such as hearing impairment?

A definitive positive answer to this question is given in the phenomenon of comodulation masking release (CMR). Hall et al (1984) demonstrated this type of release from masking initially as a paradoxical limitation to the critical band phenomenon in the spectral integration of power (Fig. 2). Generally, added masking energy on the skirts of the critical band around the tone to be detected causes the threshold to deteriorate, although negligibly if the added energy is well outside the band; indeed this is how the positions of the band's edges are defined.

However, the envelope of the added flanking noise can be contrived to have not the usual random relationship to that in the critical band, but a common amplitude envelope; this is done by co-deriving the two noise bands by a frequency-shift technique. Now detection improves with the adding of a flanking band; hence co-modulation masking release (CMR). Obviously the paradox indicates that energy considerations are no longer relevant, being superceded by co-variance considerations. The flanking band functions as the carrier of a co-variate, allowing the partialling or cancelling of the power variance of the noise. That power variance is, in a statistical decision model of detection, the major equivocating variable between the no-energy (blank) and energy (signal) trials, and hence is the main obstacle to detection. Cancelling of the equivocation is not complete, due to imperfections in the registration, estimating or discounting processes, but at between 6 and 12 dB improvement in various stimulus conditions, it represents a worthwhile enhancement. The biological value obviously lies in the ability to handle adverse signal-to-noise ratios, for example separating out interfering sounds that have modulation patterns fairly common to all the bands in their f-spectra, such as unwanted voices at cocktail parties. Warren and Verbrugge's bouncing bottle can be distinguished from their breaking bottle on the basis of degree of co-modulation - the proportion of amplitude fluctuation that is common to all the spectral bands.

The lack of any anatomical or physiological evidence for neural lateral interaction in f-space before the cochlear nucleus suggests that the CMR phenomenon is not based in the cochlea, although a role for efferents affecting the cochlea cannot be ruled out. To maximise the effectiveness of the discounting process in CMR, the auditory system should minimise cumulative introduction of neural noise; in other words the estimate of noise power variance should be cancelled at the most peripheral level possible. The cancelling process need not occur at a single level as it is a universally useful principle. This realisation renders interpretable the finding in our more recent unpublished experiments that a CMR of about 12 dB can be obtained with the flanking band on the same ear, but only about 4 dB with it on the opposite ear, under otherwise similar conditions. The cancelling process seems to be strictly range-limited in F, falling off in its efficiency quite steeply beyond 8 cycles/sec (Hall and Haggard 1983). It thus probably does not reflect the peripheral mechanism for the enhancement of the modulation spectrum of speech, which I earlier hypothesised to be required for optimal processing. That general peripheral enhancement is apparently given by short-term-adaptation, and its effects can be seen in the internal representation of modulations up to somewhat higher F-values than 8 cycles/sec, as revealed in temporal masking patterns (Plomp 1983) and in some absolute sensitivity to modulations of much higher frequency (Viemeister 1979). The range-limitation of the operations underlying CMR is not in conflict with these broader F-range requirements at the periphery. If speech and other important or interfering sounds have most of their modulation energy at low F, then much of the correlation between the actual modulation patterns in their various audio frequency bands must be carried in the gross envelope pattern, i.e. by the low-F range. The use of the high-F information in abrupt changes can proceed by parallel processes of feature extraction, either logically upstream or downstream from the stage which uses the less informative low-F variation in the CMR process.

Summary of perceptual effects

In this section I have attempted to illustrate ways of analysing experimentally the use that is made of temporal patterning, to give a richer theoretical appreciation of why its degradation, whether of central or cochlear origin, should contribute to auditory disability. As frequency-analysis is primary, the internal representation of the apparently temporal aspect of spectro-temporal patterns such as glides and formant transitions must partly suffer in hearing impairment. Speech components are grouped together on a comodulation basis, and comodulation analysis can be used effectively to enhance monaural signal-to-noise ratios. These demonstrations of the role of temporal processing yield extra appreciation of the difficulties suffered in hearing impairment of cochlear origin. We cannot yet be sure of exactly why hearing-impaired listeners' difficulties are compounded; is the extra difficulty due simply to the TMTF consequences of the loss of frequency resolution for spectro-temporal patterns; how much is secondary in a developmental sense yet independent and additively deleterious in its concurrent effects; and how much is functionally independent, but due to some common remote underlying cause such as cardiovascular disease?

Few techniques are available for tackling the important issue of in(ter)dependence of frequency- and temporal-resolution effects. One approach to the distinction between the first two possibilities would be to model the effects of degradations in frequency resolution and temporal resolution independently and then jointly, matching the numbers and patterns of errors produced (Fig. 3). The

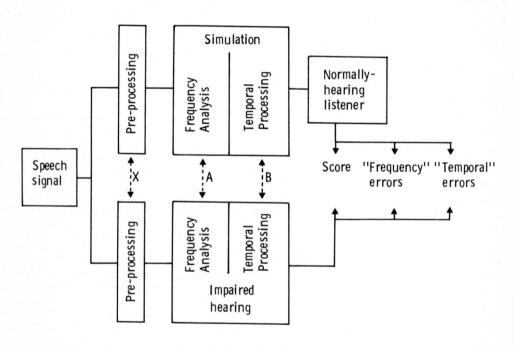

Fig. 3. A schema to represent the generation of common patterns of performance on a word identification task from impaired listeners and normally-hearing listeners experiencing a simulation of aspects of impairment. Sets of parameters (A) and (B) can be adjusted to model these output measures and others in psychoacoustic experiments; the adjustment of pre-processing parameters (X) may give beneficial effects upon these measures similar in the simulation and in the hearing-impaired listener.

simulated loss of frequency resolution will generate some errors of a "temporal" type. Separate modelling of the loss of temporal resolution in individuals would determine the excess temporal errors not explicable in this way. The simulation approach also offers a way of investigating any beneficial effects of pre-processing in relation to such internal distortions. However the optimum pre-processing may not necessarily be the inverse of the assumed distortion.

IMPLICATIONS FOR SIGNAL-PROCESSING IN THE SPECTRO-TEMPORAL DOMAIN

In this final section I wish to outline how our appreciation of temporal patterning in speech sounds and the ways of quantifying and transforming the patterning, as introduced in the first two sections, should inform our approach to processing auditory signals in telecommunications and in aids to impaired hearing.

We have already noted that Langhans and Strube (1982) used an analysis-synthesis system to modify the modulation spectrum of speech on resynthesis, without introducing other deleterious distortions. Because of the highly abstract relation between the tilfering and the actual speech waveform or spectrum the finer details of their approach will not be discussed here. But considerations of the effects of such processing cast light on the source of benefit in non-time-invariant and non-linear signal-processing generally. I will discuss some limited similarities between another less abstract system and tilfering as ways of making the modulation spectrum more nearly flat.

Spectral approaches to spectro-temporal enhancement

Flattening the modulation spectrum by tilfering has one characteristic in common with a conventional AGC amplitude compressor, that large low-F modulations are reduced. However it can also be extended to enhance lesser mid- and high-F modulations which a negative feedback AGC does not ordinarily do. To achieve something similar with gain-adjusting devices, a bank of expanders (circuits with positive rather than negative gain-feedback) would be needed. The tilfering approach is more comprehensive and specifiable, yet probably little more costly computationally. Multiband compression has been suggested (Evans 1978) as a way of matching speech signals to the ear's limited dynamic range. Applying AGC non-specifically to a large number of bands reduces spectral contrasts; the reverse would rather appear to be required, as just argued. AGC in a very few bands appears to suffice for overall matching of dynamic range across frequency (Laurence et al 1983). Solution of the dynamic range problem on a multi-band basis therefore appears partly opposed to enhancement, but if a stage of overall AGC on very few bands is made to precede high pass tilfering or expansion in sub-bands, the contradiction is resolved.

Though useful as preprocessing to resist degradation by added noise, and reportedly beneficial in hearing impairment, Langhans and Strube's work has not yet been followed by reports of systematic exploration of the effects of transformed F-spectra. Here it is necessary to draw some distinctions between possible enhancing effects in (a) pre-processing for transmission through noise, (b) pre-processing to offset the degraded frequency resolution of hearing impairment, and (c) post-processing after contamination with noise. The three possibilities are ranked in ascending order of challenge and descending order of promise for the engineer, but insofar as a technique may prove successful in (a), then applying it to (b) then (c) could be worthwhile.

Wherein could the possibility for enhancement as reported by Langhans and Strube lie? Two classes of explanation must be distinguished, even if they are assumed to have been linked in the feedback process of biological evolution. These are again the physiological and the ecological. Any account must acknowledge that auditory analysers have a limited capacity to process amplitude-change information as a function of time. A physiologically realisable way of handling this limitation would be, by analogy with f-space, to have each "channel" convey the pattern or amount of modulation for only a limited range of F-values. Could an overloaded channel then be relieved by tilfering out the uninformative low-F changes, giving a type of release from F-masking? There is as yet no direct evidence for any such F-masking effect to be relieved by F-range limitation as such. Dynamic range limitation for modulations in speech can be parsimoniously explained by the conventional spectral concepts of threshold and discomfort level. Amplitude compression has negligible effects upon speech intelligibility in quiet, and its beneficial effects as a preprocessor before noise, (application (a) above), are adequately explained by the relative amplification of weak sounds to avoid conventional f-masking, Haggard (1983). Thus the physiological account, linked to the earlier-mentioned idea of F-analysers of gradated centre-F, and tuned to different sub-ranges of F, seems implausible.

The view more likely to turn out correct is an ecological one, justified by analogy with the telecommunication engineer's attitude to f-bandwidth. The available bandwidth is ideally centred on the frequency region of greatest information content; the energy is then redistributed therein to lift as many as possible of the equal importance-weighted bands above a critical S/N ratio. In F-space this amounts to removing fluctuations of low F, (which AGC compression also does, making little difference to word intelligibility), and flattening the modulation spectrum, at least over an F-scale weighted according to F-importance values. This extension of the analogy to F-space is not necessarily based on the assumption that some external noise to be encountered has a modulation spectrum that is flat. That form of the analogy would lead us back to the postulate of tuned F-analysers. The more appropriate analogy with audio engineering is to internal quantisation noise in digitisation with few quantisation levels; preflattening the spectrum improves the f-importance-weighted S/N ratio in the re-converted analogy signal. A limited internal amplitude-handling capacity is therefore the probable explanation of Langhans and Strube's results. The emergence of the effect as enhancement against added noise is consistent with the continuous noise using up the dynamic range even for components that may not be masked or only partially masked in the conventional sense. This view is in turn consistent with the CMR phenomenon.

Some implications of enhancing intensity resolution in time, by scaling up the high-F modulations, can be explored through an example. Let us define the fundamental perceptual problem as that of judging whether or not a speech formant has moved in a particular direction as in Figure 4, over a certain fixed time (t_2-t_1). In the

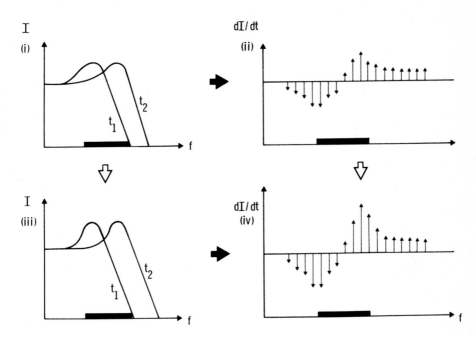

Fig. 4. The partial equivalence of enhancement in the spectral and temporal domains. Horizontal transformations are sensory, vertical ones artificial. The upper path from panel (i) to panel (iv) involves a reversal of the possible sequence of transformations for the sake of compactness. For details see text.

case represented by a sequence of conventional spectrographic sections (panel i) intensity increments and decrements of moderate value are registered over two fairly wide frequency bands. If these increments are obtained by an internal coding process of single time-differentiation (ii), and enhanced by a further external stage of differentiation, an effect more like (iv) will result, more extreme in magnitude but redistributed in narrower frequency space. This is because only down on the skirts of the formants will change be fast enough to benefit from the time-differentiation of spectral amplitude entailed in high-pass tilfering. The effect of the pre-processing will depend upon the f-integration characteristic for change. It is by no means certain that auditory resolution, the formant bandwidths of speech, and the velocities of articulatory movements are already mutually optimised for intelligibility. However, in our chosen dynamic paradigm the effect of the simple form of tilfering by introducing sequential differencing turns out to be similar (panel iv) to that of narrowing the formant bandwidths by operating on successive samples purely in the spectral domain (panel iii). Thus there are some contexts in which the distinction between loss of frequency resolution and loss of time resolution is not worth making. Of course the two approaches will give different behaviour for static frequencies with abrupt onsets and offsets. This near-equivalence suggests that progress towards automatic speech enhancement will be like tacking between Scylla and Charybdis against a headwind. Parameters of processing will need to be optimised in relation to particular asumptions about subsequent distortions to be suffered, and may need to be different for different classes of speech sound.

Enhancement of spectro-temporal contrasts can be achieved either by bandwidth-narrowing (as in Fig. 4) or by making formants rect-angular in the spectrum, i.e. flat-topped with steep slopes. Any method that starts with a spectral representation in terms of fairly narrow frequency bands has the option of considering many different form of subsequent temporal processing; but in practice it may be computationally necessary to accept more limited possibil-ities by describing the short-term spectrum with higher order parameters. Sidwell and Summerfield (1984) have performed a number of experiments on schemes of the bandwidth-narrowing type and have reviewed others, oriented towards signal-processing for hearing impairment; broadly speaking, no improvements in intelligibility have resulted. The alternative is even less promising. Early work in speech synthesis used control-signals that were in effect quantised in spectral amplitude (Cooper et al 1951). This flattened the tops and steepened the sides of formants; the abrupt level changes in individuals harmonics gave "glouglou" noises and pro-vided intelligibility only good enough for exploring use of spec-tral cues to phonetic distinctions in somewhat practised listeners. Thus attempting to achieve spectral contrast enhancement, and there-by to achieve some of the postulated desirable temporal enhance-ment, does not appear to be very promising at present.

From the foregoing we may conclude that optimum forms of spectro-temporal pre-processing are likely to be quite abstract, different from individual to individual, and may even have to vary over dif-ferent frequency bands and over time, because they must depend upon matching sound properties to both time- and frequency-resolution. But if the f-integration characteristic for registering changes of intensity, represented by the thickened bar on the x-axis in Fig. 4 were known, a more decisive step could be taken towards tailoring formant representations optimally. It is unlikely that we can achieve enhancement of spectro-temporal patterning by the con-ceptually simple approaches that operate purely in the spectral or cepstral domains. Any optima are likely to be located in relatively modest plateaux in the abstract space defined by parameters of signal processing.

Against the limited success achieved so far must be considered the listener's great familiarity with unprocessed speech and the hearing-impaired person's familiarity with his own distorted trans-form of it in particular. Improving on Nature is a tall order, but with pre-processing that makes speech intelligibility at least no worse for the normal ear there are grounds for cautious optimism and experiments on perceptual re-learning.

Temporal approaches to spectro-temporal enhancement

The most general and powerful approach to processing spectro-temporal patterns would be 2-dimensional high-pass filtering to give enhancement in frequency and time simultaneously. This suffers from a double disadvantage: (i) despite the great advances in measurement obtained through spatialising time in the spectrogram, from the point of view of the auditory system, time and frequency are made of different stuff. Suitable transformations of each to enable parameters of the analysis to be shared for the sake of computational economy might be very hard to find; (ii) the computational burden is massive. We are left for the present to consider the implications of primarily temporal schemes.

Tilfering and multi-band compression/expansion have already been discussed in some detail in order to expound the general argument. In brief, tilfering appears to offer a wider range of possibilities, being more general. From the point of view of economy of software and hardware some similar effects could be reached by having bands of automatic gain controls in series with expanders. However, as with other types of scheme, the best strategy seems to be to use powerful and general techniques first to find any potential benefits then later to worry about economic realisation.

One possible realisation that has much in common with high-pass tilfering is adaptive convergent filtering as shown in Fig. 5. This has been pointed out by my colleague JR Trinder who has prepared the demonstrations of Fig. 6, in the course of implementing adaptive filtering for research purposes, and has documented the responses of the filter coefficients to various step-function and

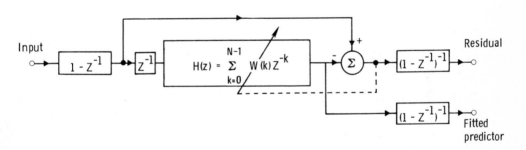

Fig. 5. Schematic block diagram of the least-mean-square Widrow-Hoff adaptive filter. The fit of the N-coefficient predictor of the speech wave is updated (dotted line) to minimise the mean-square error. Both predictor (fit) and error (residual) are available as outputs. H(z) represents the transfer function of the fitted predictor expressed in terms of the z-transform and the N-coefficients.
$(1-Z^{-1})$ and $(1-Z^{-1})^{-1}$ respectively represent pre-equalisation (+6 dB/octave) and post-equalisation (-6 dB/octave). The post-equalisation is necessary for speech signals to prevent overall flattening of the speech spectrum. Pre-equalisation is optional but leads to more efficient use of the adaptive filter.

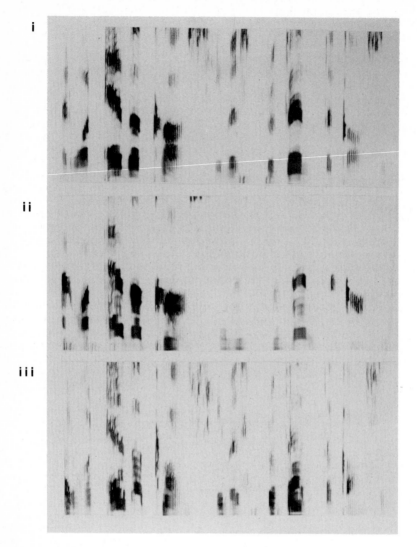

Fig. 6. Speech spectograms (0–4 kHz) of the same utterance subjected to double adaptive convergent filtering giving 12 dB/octave tilfering of the modulation spectrum. (i) Original sentence: "At whatever cost, he must prevent the gossip" (ii) low pass, cut-off at 4 cycles/sec (iii) high pass, cut-off at 32 cycles/sec. In each case the transfer function is modelled with 64 coefficients (effectively 32 resonant filters up to 5 kHz, whose parameters are continually updated). Versions (ii) and (iii) show respectively loss of detail and loss of spectral dynamics. The pattern-extraction processes of visual and auditory systems both accept the high-pass version as more like the original than the low-pass version; it has the higher information rate. Time scale: 1.8 sec.

modulated inputs. The technique is typically used to fit a spectral model to the quasi-stationary spectral properties of a signal, enabling the dissociation of the parameters of this fit from the quasi-dynamic residual. That general approach has two almost opposed applications in speech. They are distinguished by where one locates the border between quasi-stationary and quasi-dynamic signals. In linear predictive coding (LPC: Atal and Hanauer 1971) the fit is made sufficiently frequently (typically every 10 ms) to sample the movement of the spectral phonetically-relevant parameters of speech; the quasi-stationary fit parameters are transmitted and the spectral residual is in general discarded. LPC is not essentially temporal but part of the residual is due to spectral effects of source characteristics which tend to change from frame to frame. For speech enhancement in the presence of quasi-stationary noise (Widrow et al 1975) a continuously changing fit is made to the static or slowly-changing properties of the signal-plus-noise spectrum. This is then discarded and the high-f-pass residual signal is transmitted; now the failures of past spectra to predict the present spectrum, i.e. the spectrotemporal changes, are emphasised. (There is a further difference in that LPC is performed on a discontinuous frame-by-frame basis, whereas adaptive filtering is performed continuously by taking in successive waveform samples). Both LPC and adaptive filtering acknowledge that the mid-F components are important for speech intelligibility. The time constant of convergence in the adaptive process can be set to put more or less weight on past relative to present waveform samples in determining the fit, and this directly affects the F cut-off. According to where the cut-off is placed and other circumstances it may be more advantageous to transmit the economical (low-F-pass) fit (as in LPC) or the (high-F-pass) residual (as in most enhancement applications of adaptive filtering). For a single convergence rate (i.e. single F of low- or high-pass cut-off), auditorily comparing the fit to the residual does not yield a striking subjective difference. Two-stage processing (12 dB/octave) and differing cut-offs are required for clearly appreciable differences to emerge (Fig. 6).

Obviously adaptive filtering in its application to post-processing after noise has been added will be of limited help where the interference has a broad f-spectrum and a similar F-spectrum to that of the speech; the interference must be relatively static to be rejected. Irrespective of the F-spectrum, its major successes are met when the (quasi-static) f-spectra are harmonic, yet, highly atypically for speech, are also monotone; the residual can be easier to listen to when the original contains reverberation, engine noise, jet whine, fan hum and some types of music.

Our applications here are different, not being concerned with post-processing but with simulation and compensatory pre-processing. The spectrograms in Figure 6 and the associated auditory demonstrations show the difference between the adaptive-filter approximations to low-pass tilfering and to high-pass tilfering, in the fit and the residual outputs respectively. The former is not a true simulation of a hearing impairment, nor even a fully adequate simulation of its temporal aspects, as the parameters of temporal blurring have not been adjusted to match psychophysical data. It serves however as a general indication of the effects of blurred temporal resolution upon speech features. The fitted predictor has an audible fluctuation of its spectral tilt, rather like selective fading in high frequency radio transmissions, and has a slightly muffled quality. The residual sounds harsh and has reduced harmonicity but near-natural clarity, indicating that less of phonetic value has been discarded in this case.

The residual after adaptive filtering creates temporal enhancement, in a more general and essential way than the spectrally-based methods listed in the previous section. The rate at which the filter fit converges will determine, in effect, only the cut-off in the TMTF, while the number of passes determines the slope. This constrains the possible ways in which the signal can be tilfered. These constraints are, of course, the price of computational economy by comparison with Langhans and Strube's more flexible method. Also the process does not permit any material modulation gain at any F without extra provision. Given that high-pass tilfering the F-spectrum by +6 dB or +12 dB per octave is ex hypothesi what is required, and the low-F high amplitude variations seem to be uninformative, the limited tilfering options in the adaptive filtering method may be acceptable.

Finally we may note that linear and non-linear contraction and expansion of the frequency scale have already been tried in signal processing for hearing-impairment, bandwidth compression, and correcting helium speech. The same untying of time from frequency in analysis-synthesis or in sample-and-discard processing also permits multiplicative or non-linear warping of the time scale. There is tentative evidence that some hearing-impaired listeners profit from cue-enhancement in the time domain i. e. the exaggeration of duration distinctions (Revoile et al 1984). The next step would be spinning out in time portions of fast change or intervals with which differential discrimination difficulties are experienced. There is enough evidence (Bailey 1983) for temporal discrimination difficulties relevant to speech, beyond difficulties with absolute detection of 10 ms gaps, for this to be worth further investigation. Obviously to maintain real time operation, time-expansions have to be at the expense of time-compression of steady-states where those steady-states are least informative and a material mean delay is inevitable. Such processing would be computationally complex because of the need to make phonetically wise decisions about what to discard so as to leave some real time for elongations, but this is well within the reach of present-day concepts and techniques. One danger is that phonetically meaningful ratios of durations would be disrupted, but these relativities might be subject to some relearning, provided that the processing could transform them in a consistent fashion.

In this section I have considered various ways in which we might consider enhancing the temporal structure of speech sounds, particularly those ways to which the concept of the modulation transfer function gives us access. No miracles have yet emerged but there are grounds for cautious optimism.

CONCLUDING SUMMARY

We have seen that the modulation spectrum and the TMTF offer at least a statistical way of describing the spectro-temporal pattern information in speech. There is experimental evidence to suggest that the perceptual use of spectrotemporal information is basic to important perceptual processes of grouping when more than one source or interfering noise may be competing for attention. Deficits of temporal processing contribute to auditory disability, but their pathophysiological and perceptual status in relation to primary frequency resolution deficits remains unclear. Sophisticated digital signal-processing is becoming accessible enough to consider the possibility of improving intelligibility and quality of speech through enhancing the statistically informative contrasts in spectral and temporal domains, or decreasing the less informative contrasts, or both. The TMTF offers one way of describing the effects of temporal degradations and enhancements, which are otherwise difficult to specify and conceptualise. The spectro-temporal patterning of speech entails that spectral and temporal sharpening will sometimes be rather similar in their effects. A demonstration that enhanced intensity contrast in time within speech signals can specifically and materially offset difficulties of listeners with deteriorated temporal resolution has yet to be made.

ACKNOWLEDGEMENTS

I thank Julian Trinder for helpful discussions of adaptive filtering and the demonstrations in Figure 6.

REFERENCES

Atal BS, Hanauer SL (1971) Speech analysis and synthesis by linear prediction of the speech wave. J Acoust Soc Amer 50:637-655
Bailey P (1983) Hearing for speech. In ME Lutman and MP Haggard (Eds.) Hearing Science and Hearing Disorders. London: Academic Press, pp 23-31
Buus S, Florentine M (1984) Gap detection in normal and impaired listeners: the effect of level and frequency. This volume
Cooper FS, Liberman AM, Borst JM (1951) The inter-conversion of audible and visible patterns as a basis for research in the perception of speech. Proc Nat Acad Sci 37:318-325
Darwin CJ, Sutherland NS (1984) Grouping frequency components of vowels. Quart J Exptl Psychol 36A:193-208
Delgutte B, Kiang NY-S (1984) Speech coding in the auditory nerve: IV. Sounds with consonant-like dynamic characteristics. J Acoust Soc Amer 75:897-907
Evans EF (1978) Peripheral auditory processing in normal and abnormal ears. Scand Audiol Suppl 6:9-48

Evans EF (1984) Aspects of the neural coding of time in mammalian peripheral system relative to temporal resolution. (This volume)

Haggard MP (1983) New and old conceptions of hearing aids. In ME Lutman and MP Haggard (Eds.) Hearing Science and Hearing Disorders. London: Academic Press

Hall JW, Haggard MP (1983) Co-modulation - a principle for auditory pattern analysis in speech. Proc 11 ICA 4:69-71

Hall JW, Haggard MP, Fernandes M (1984) Detection in noise by spectro-temporal pattern analysis. J Acoust Soc Amer 76:50-57

Harrison R, Evans EF (1979) Some aspects of temporal coding by single cochlear fibres from regions of cochlear hair cell degeneration in the guinea-pig. Arch Otorhinolaryngol 224:71-78

Henderson D, Salvi R, Pavek G, Hamernik R (1984) Amplitude modulation thresholds in chinchillas with high-frequency hearing loss. J Acoust Soc Amer 75:1177-1183

Holmes JN (1961) Some measurements of the spectra of vocoder channel signals. Research Report 20651, UK Post Office Dollis Hill Research Station

Horner K, Bock GR (1984) Electrically-evoked unit responses in the inferior colliculus. Develop Brain Res (In Press)

Kelly PH (1971) Theory of flicker and transient responses. J Opt Soc Amer 61:537-546

Lackner JR, Teuber H-L (1973) Alterations in auditory fusion thresholds after cerebral injury in man. Neuropsychologia 11:409-416

Langhans T, Strube HW (1982) Speech enhancement by non-linear multi-band envelope filtering. Proc IEEE 5th ICASSP, pp 156-159

Laurence R, Moore BCJ, Glasberg B (1983) A comparison of behind-the-ear high-fidelity linear aids and two-channel compression aids in the laboratory and in everyday life. Brit J Audiol 17:31-48

Lundin FJ (1982) The influence of room-reverberation on speech. Speech Transmission Laboratory QSR 2/3, Royal Institute of Technology, Stockholm pp 24-59

McAdams S (1984) The auditory image. In WR Crozier and AJ Chapman (Eds.) Cognitive processes in the perception of art. Amsterdam: North Holland

Plomp R (1983) On the role of modulation in hearing. In R Klinke and R Hartmann (Eds.) Hearing-Physiological Bases and Psychophysics. Heidelberg: Springer Verlag

Revoile SG, Holden-Pitt LD, Pickett JM (1984) The effects of duration adjustments of preceding vowels on fricative voicing perception by hearing-impaired listeners. J Acoust Soc Amer 75:(Suppl 1) S5

Schroeder M (1981) Modulation transfer functions: definition and measurement. Acustica 49:179-182

Shailer MJ, Moore BCJ (1983) Gap detection as a function of frequency bandwidth and level. J Acoust Soc Amer 74:467-473

Sidwell A, Summerfield AQ (1984) The effect of enhanced spectral contrast on the internal representation of vowel-shaped noise. J Acoust Soc Amer (Submitted)

Steeneken HJM, Houtgast T (1980) A physical method for measuring speech-transmission quality. J Acoust Soc Amer 68:318-326

Steeneken HJM, Houtgast T (1983) The temporal envelope spectrum of speech and its significance in room acoustics. Proc 11 ICA 7:85-88

Summerfield AQ, Foster JR, Haggard MP, Gray S (1983) Perceiving vowels from uniform spectra. Percept Psychophysics 35:203-213

Tansley BW, Suffield JB (1983) Time course of adaptation and recovery of channels selectively sensitive to frequency and amplitude modulation. J Acoust Soc Amer 74:765-775

Tyler R, Summerfield AQ, Wood E, Fernandes M (1982) Psychoacoustic and phonetic temporal processing in normal and hearing-impaired listeners. J Acoust Soc Amer 72:740-752

Viemeister NF (1979) Temporal modulation transfer functions based upon modulation thresholds. J Acoust Soc Amer 66:1364-1380

Viemeister NF (1980) Adaptation of masking. In Gvd Brink and EA Bilsen (Eds.) Psychophysical, physiological and behavioural studies in hearing. Delft: Delft University Press, pp 190-198

Warren WH, Verbrugge R (1984) Auditory information for breaking and bouncing events. J Exptl Psychol (HP & P): In Press

Widrow B, Glover JR, McCool JM, Kaunitz J, Williams CS, Hearn RH, Zeidler JR, Dong E, Goodlin RC (1975) Adaptive noise cancelling: principles and applications. Proc IEEE 63:1692-1716

Wilson JP (1970) An auditory after-image. In R Plomp and GF Smoorenberg (Eds.) Frequency analysis and periodicity detection in hearing. Leiden: AW Sijthoff, pp 303-315

Subject Index

A short guide to the major topics discussed in the different chapters is given in the Introduction. Abbreviations are explained in the Index, and a reference is given to pages where specific terms are defined.

accuracy 44,96-97,182-184
acoustic biotope 104
adaptation 41-43,84,221-223
adaptive filtering 231-234
acf = auto-correlation
    function
AGC = automatic gain
    control 227
aggression song 24
aids to impaired hearing
    227-234
annoyance 205
anurans 58-71
ambiguity 102
amphibian papilla 59
amplitude compression
    227-228
amplitude modulation 4-6,
    16-25,37,40-44,66-69,87,
    109,113-116,135,146-147,
    216-227,231-234
analyser of modulation fre-
    quency 220-221
a-priori information 102
artificial song 5-6,9
attenuation mechanism 199
attractiveness 9
auditory: see next word
autocorrelation 51,53,80,102,
    110,118,187

backward masking 132,152-153
band-pass neuron (temporal)
    22-23,41-43,65-70
bandwidth 30,75-83,133-134,
    144,160-162,165,175,190-192
basilar membrane 111-112,
    206-207
basilar papilla 59
bat 180-204
BD = best delay 197
behavioural test 20
behaviourally relevant signal
    104
best frequency 111,208-209
binaural image 137
binaural pitch effect 112,118

binaural time constant 137
bird 13,36-37,104,112-119
BMF = best modulatory frequency
    112-116
brain 21
bug 4
bush cricket 4,25

calling song 21
carrier frequency (effect of)
    20
cascade of resonators 144
cat 76-78,83,89,114,116-118
cavefish, blind 30-31
ccf = cross-correlation function
cepstrum 216
cerebral cortex 196-198,222
CF = characteristic frequency
cf = constant frequency
cf-signals (bats) 180
characteristic frequency 75,89
chinchilla 83,86,160
chirp 4-6,22
chopper response 114
cicada 16
click (train) 15,62
CMR = comodulation masking
    release
cochlear filter 83,111-112
cochlear nerve: see nerve,
    auditory
cochlear nucleus 87-91,114,194
cochlear pathology: see pathol-
    ogical conditions
cod 30-31,34
coding, temporal 25,41-43,
    66-69,108-121
coefficient of: see next word
coherent optimal receiver 188
coincidence effect 114-116
communication theory 99
comodulation masking release
    224-225
compression of amplitude 227-228
contrast 87
cortex, auditory 196
cos+/cos- noise 50-54

courtship song   24
cricket   4,13-14,20-25
critical band (width)   21,
    143,148,161-163,220,224
critical duration   124-126,
    134
critical masking ratio
    30-31
cross-correlation   13,
    113-116,118,187
cycle histogram   61-65

damping, resistive   15
d' (detection index or sen-
    sitivity of energy
    detector)125,133-134,161,
    164-165
decay of auditory activity
    (sensation)   43,130-132,
    149-150,160,168
decay, oscillatory   13,75-81
decrement detection   38-40,
    132,150-155
delay discrimination   52
delay element   100,144
delay sensitive neurons   196
Delta function   11-14
deterministic signal   43
Dirac's Delta function: see
    Delta function
DL = difference limen   30
don't care case   105
dynamic range   25

echolocation   180-204
EF = exciting frequency
    196
effective bandwidth   75
eighth nerve: see nerve,
    auditory
electric spark   12
energy detector   132-135,
    160-163
energy splatter: see
    spectral splatter
enhancement (of speech)
    228-234
envelope detector   98
envelope resolution   33-34,
    47-48
epsp (excitatory post
    synaptic potential)   41-42
equivalent rectangular
    band-width   75

false alarm   102

filter, auditory   10-15,30,
    111-112,142,144,160,168
filter-bank model   142
finite impulse response   100
FIR = finite impulse response
fishes   28-57
flatfish   30-31
fm = frequency-modulated
fm-signals (bats)   180
formants   111,218,221,223,229
forward masking   37,92-93,111,
    129,132,145,148,150-151
Fourier-transform   13,108
frequency (definition)   142,216
frequency analysis (in ears)   4,10,30
frequency cues (in fish)   46-48
frequency (effect of)   20,34,
    47,163-171
frequency filter   9
frequency modulation   9,87,
    180,194-198
frequency selectivity (dis-
    crimination)   30-31,44,52,
    151,159,207
frequency sweep   8
frequency threshold (tuning) curve
    (FTC)   30-33,58,75,79,144
frog   13,58-71,104-105
FTC = frequency threshold curve

gap detection   4-6,37-44,92,
    130-135,152-154,159-177,
    221-222
gerbil   87
goldfish   28-54
grasshopper   4-16
guinea fowl   104,112-119
guinea pig   78,83,87

hair cells, orientation of 28-29
harmonic   79,111,197-198,223
harmonic ratio   110
harmonic relation   114
harmony   117
hearing damage, risk   205-214
hearing disorders: see pathol-
    ogical conditions
hearing impairment: see impaired
    hearing
high-pass neuron (temporal)
    22-23,65-66
high-resolution method   104
Huffman sequences   128,205
human psychoacoustics   37,40-41,
    48-51,122-138,159-177,215-235
human tympanum   13

ideal energy detector 133-134,
   161-163
impaired hearing 37,83,92-93,
   154,159-160,170-177,205-214
impulse (in song) 5
impulse response 12-15,75-83
impulsive stimuli 80,207-214
increment detection 38-41,
   132,150-155
inefficiency parameter 134
industrial noise 210-214
inferior colliculus 116,194
information processing 122
inhibition (neural) 87
inner ear (anuran) 59
inner ear (fish) 28-30
insect 3-26,183-184
integration, temporal 33-37,
   40,43,124-126,133-134,
   141-142,145-148,159-163,
   175,205-206
intelligibility (of speech)
   218-220
intensity: see level
interaural correlation 237
interspike interval 53,60,
   81,111
interval discrimination (for
   bursts) 46
intrinsic oscillation
   114-116,118
ISI = inter-spike-interval

jitter (temporal) 45-46,
   185-186

lagena 28-29
larynx 219
laser vibrometry 11
latency-constant neurons 195
lateral image 137
lateral inhibition 101
leaky integrator 141,145
level (of sound), effect of
   6,18,34,80,85,87,92,
   131-132,162-171,208-209
level, discrimination of
   27,29,150,154
L_{eq} = equivalent continuous
   sound level 212
linearity 143,208
linear predictive coding 233
lizard 13
LPC = linear predictive
   coding
locust 10,14-16
loudness 205-214

low-pass neuron (temporal)
   22-23,65-66,89

macula 28-29
mammals 13,36-37,41,74-93,
   114-118,160,180-204,222
masking 31,34,85,92,129,
   132,145,150-151,160,
   170-171
masking release 224-225
matched filter 102,187
mating calls 4,21,24,59,
   69-71
MDG = minimum detectable gap
   duration (see gap detection)
membrane potential, decay
   of 22
midbrain 64-71, 112,115
middle ear 199,208-209
MLD-nucleus (bird) 113
missing fundamental 108
mode of vibration 15
model 11,81,97-104,
   142,189-191
model for gap detection
   132-135,160-163
modulation of frequency
   9,87,180,194-198
modulation of amplitude: see
   amplitude modulation
modulation spectrum 216-235
mole cricket 4
moth 11-12,16-20
mouse 222
multiband compression 227
multiplication (neuronal)
   114-115

nerve (fibers), auditory
   31,34-53,58-64,74-93,112
neural inhibition 87
non-harmonic signals 118
non-linearity 16,41
non-recursive filter 100
noise (external) 20,30,34,
   36,105,205-214,228,233
noise (internal) 101,
   161-163,225
noise power variance (in
   speech) 224-225

off-frequency listening 40
off-inhibition 87-91
off-suppression 84-93
optimal receiver 187-191
optimum statistical detector
   134

order, temporal (discrimation of) 128-129
oscillatory decay 13,75-81
otolithic ear (fishes) 28-55
outer ear 208-209

pathological bandwidth 81
pathological conditions 37,
  83,145,171-177,215,222
pattern perception 104,
  110-111,221-227
periodicity coding 47,60,220
periodicity equation 116
periodicity pitch 49,108-121,
  216
peripheral filtering 30,142,
  144,160,168
phase information 188
phase locking 20,28-29,
  44-46,61-65,79,92,111
phase spectrum 127
phonetic segments 219
phonotactic response 20,23
pitch 49-52,136,142,
  108-121,216
place principle 108,111,118
plainchant 218
power function 34
power spectrum 127
pressure difference
  receiver 13
probe, temporal 148-149,
  152
propagation delay 101
prothoracic ganglion 21
PST (post-stimulus-time)
  histogram 38-39,84-88
psychoacoustics 28-54,
  91-92,122-138,159-177,
  205-214,215-235
psychophysical tuning curve
  31,145,151
pulse compression 187

Q = tuning factor 67,209

range resolution 182-184,
  193-194
range sensitive neurons 196
rat 78
RC network 207
recovery 43,85,91
redundancy 6
refractory period 53
release of neurotransmitter
  41-43
repetition noise 49-53

residual pitch 108-110,117
resistive damping 15
resolution, temporal 11,
  17-22,33-34,44-46,81,87,
  96-97,123,126-134,159-178
resonance 143-144,207
response saturation 24
response song 9
reverberations 218
reverse correlation 75-77
rhythmicity 3-4
ringing 160-161,175
rippled spectrum 136
rms (root mean square)
  duration 129
roughness 148

sacculus 28-55
SAM = sinusoidal amplitude
  modulation (see
  amplitude modulation)
saturation of response 24
second order system 13,143-144
selectivity, temporal
  22-23,41-43,65-70
semicoherent receiver
  188-192
signal-to-noise ratio 99,
  102,133,189-191,226
simple resonator 13,143-144
simulated impairments 160,
  170-177,233
S/N = signal-to-noise ratio
social signal 4,21,24,59,
  69-71
soldierfish 30-31
sonar 180
sound level: see level
sound level meter 205-214
spatial acuity (in vision)
  126-127
spectral (energy) splatter
  40,153,171
spectral whitening 100
speech, perception of
  215-235
speech, synthesis of 230
spike interval: see interspike interval
spike rate/intensity function 36,111
splatter: see spectral
  splatter
spontaneous activity 83,92
squirrelfish 30-31
statistical detection 134,146
stochastic signals 43

stridulation 4,7-8
subharmonic 110,117
summation (temporal): see
   integration
suppression, two-tone 25
swimbladder 29-30
syllable 4-7,20-22
synaptic delay 100,114
synaptic model 41-43
synchronization (coef-
   ficient) 16-20,40,63
synchrony-suppression 81
system identification 105

teleost 28
temperature, effect of
   68-71
temporal: see next word
threshold detection 34-35,
   146
threshold (displacement
   amplitude) 29
tilfering 216,223,227,233
time constant 9,13,35-36,
   85,89,117,124-126,133-138,
   144-147,161-163,205-214
time cue 11
time delay hypothesis
   181-186
time-intensity trade 33,
   123-126,133
time locked neurons 194
time-reversed stimuli 128
TMTF = temporal modulation
   threshold function (see
   amplitude modulation)
toads 58-71
tooth impact frequency 13
topographic organization
   116
torus semicircularis 64-71
tracking neurons 196
transfer function 13,
   207-208
transfer function (temporal):
   see amplitude modulation
treefrog 62
tuning curve (frequency):
   see frequency threshold
   curve
tuning curve (temporal):
   see amplitude modulation
two-click procedure 129
two-tone suppression 25
tympanum 10-15

uncertainty principle 99

utriculus 28-29

vector strength 78-79
vibrational mode 15
vibrometry (laser) 11
visual acuity 127
volley principle 116
vocal cords 59
vowel 111,223

waveform analysis 34
Weberian ossicles 29-30
Weber's law 30,43,163
weighting curve 205
whitening filter 100